现代化学专著系列

含硫香料化学

孙宝国 著

科学出版社

北京

内 容 简 介

本书是一部关于含硫香料化学的专著,对 270 多种含硫香料化合物的发现、香味特征、制备方法、安全性、建议应用领域及建议用量等内容进行了论述,涉及大部分含硫香料的结构类型。

本书可供从事香料化学、食品化学、有机化学、应用化学、日用化学、烟草化学等研究工作的专业人员参考。

图书在版编目(CIP)数据

含硫香料化学/孙宝国著. —北京:科学出版社,2007
（现代化学专著系列）
ISBN 978-7-03-018428-3

Ⅰ.含… Ⅱ.孙… Ⅲ.香料,含硫-工业化学 Ⅳ.TQ651

中国版本图书馆 CIP 数据核字(2007)第 004086 号

责任编辑:周巧龙 吴伶伶 / 责任校对:钟 洋
责任印制:张 伟 / 封面设计:铭轩堂

科学出版社 出版
北京东黄城根北街 16 号
邮政编码:100717
http://www.sciencep.com

北京凌奇印刷有限责任公司 印刷
科学出版社发行 各地新华书店经销

*

2007 年 3 月第 一 版　　开本:B5(720×1000)
2017 年 3 月第二次印刷　　印张:12 3/4
字数:236 000
POD定价: 80.00元
(如有印装质量问题,我社负责调换)

前　言

含硫香料化学是香料化学的一个新兴的重要分支,最近40年得到快速发展,到2005年,有FEMA号的香料计2253种,其中含硫香料有270多种,占12%。在应用领域,含硫香料是咸味食品香精必不可少的关键原料,在食品香料中占有十分重要的地位。

作者所在的北京工商大学(原北京轻工业学院)从1981年开始进行含硫香料的研究工作。作者本人从1986年起一直从事含硫香料和咸味食品香精的研究工作,在该研究领域中先后承担过4项国家科技攻关项目、2项国家自然科学基金项目和2项省部级基金项目,先后获得国家技术发明二等奖1项、国家科学技术进步二等奖2项、发明专利9项,发表研究论文180余篇。本书在某种程度上是北京工商大学25年含硫香料研究工作的总结,书中的某些研究内容是首次公开出版的。在此,谨向25年来所有在北京工商大学从事过含硫香料研究的老师和同学们表示衷心的感谢!

在本书的写作期间,恰逢作者在北京航空航天大学挂职担任校长助理,北京航空航天大学浓厚的学术气氛与和谐的人文环境使作者能够在工作之余静心完成书稿,借此机会向北京航空航天大学的各位领导和同事们表示感谢!

北京工商大学梁梦兰教授审阅了全部书稿并提出了宝贵的修改意见,在此表示诚挚的谢意!

本书的出版得到了国家自然科学基金和北京市"新世纪百千万人才工程"的经费支持,谨此致谢!

由于作者学识和水平的限制,书中错误在所难免,敬请各位专家和读者批评指教。

孙宝国

目　录

前言
第一章　绪论 ··· (1)
　　第一节　含硫香料的历史 ··· (1)
　　第二节　含硫香料的特点 ··· (2)
　　第三节　中国的含硫香料工业 ··· (15)
　　参考文献 ·· (17)
第二章　含硫化合物分子结构与香味的关系 ······························ (20)
　　第一节　肉香味含硫化合物分子结构与香味的关系 ················· (20)
　　第二节　葱蒜香味含硫化合物分子结构与香味的关系 ·············· (35)
　　第三节　咖啡香味含硫化合物分子结构与香味的关系 ·············· (37)
　　参考文献 ·· (38)
第三章　硫醇类香料 ··· (41)
　　第一节　硫醇类香料的一般合成方法 ································· (41)
　　第二节　3-呋喃硫醇类香料 ··· (44)
　　第三节　糠硫醇和烯丙硫醇 ·· (49)
　　第四节　不含其他官能团的单硫醇类香料 ··························· (50)
　　第五节　β-羟基硫醇类香料 ······································ (50)
　　第六节　β-烷氧基硫醇类香料 ··································· (54)
　　第七节　γ-羟基硫醇类香料 ···································· (55)
　　第八节　β-羰基硫醇类香料 ······································ (57)
　　第九节　γ-羰基硫醇类香料 ····································· (58)
　　第十节　α,β-二硫醇类香料 ································· (60)
　　第十一节　α,ω-二硫醇类香料 ···························· (63)
　　第十二节　硫酚类香料 ·· (64)
　　第十三节　含六元芳香杂环的硫醇类香料 ··························· (64)
　　参考文献 ·· (65)
第四章　硫醚类香料 ··· (71)
　　第一节　硫醚类香料的一般合成方法 ································· (71)
　　第二节　简单对称硫醚类香料 ··· (72)
　　第三节　简单不对称硫醚类香料 ······································ (73)

第四节　糠基硫醚类香料 …………………………………………… (73)
第五节　3-呋喃硫醚类香料 ………………………………………… (75)
第六节　2-烷硫基吡啶类香料 ……………………………………… (77)
第七节　2-烷硫基嘧啶类香料 ……………………………………… (78)
第八节　β-羟基硫醚类香料 ……………………………………… (79)
第九节　β-烷氧基硫醚类香料 …………………………………… (82)
第十节　β-巯基硫醚类香料 ……………………………………… (82)
第十一节　β-羰基硫醚类香料 …………………………………… (84)
第十二节　γ-羟基硫醚类香料 …………………………………… (86)
第十三节　γ-羰基硫醚类香料 …………………………………… (87)
第十四节　β,γ-二羰基硫醚类香料 ……………………………… (89)
第十五节　3-烷硫基酯类香料 ……………………………………… (90)
第十六节　1,4-二噻烷类香料 ……………………………………… (90)
第十七节　其他硫醚类香料 ………………………………………… (93)
参考文献 …………………………………………………………………… (94)

第五章　二硫醚类香料 ……………………………………………………… (105)
第一节　二硫醚的一般合成方法 …………………………………… (105)
第二节　对称二硫醚类香料 ………………………………………… (108)
第三节　不对称二硫醚类香料 ……………………………………… (109)
第四节　环二硫醚类香料 …………………………………………… (112)
参考文献 …………………………………………………………………… (113)

第六章　多硫醚类香料 ……………………………………………………… (118)
第一节　多硫醚的一般合成方法 …………………………………… (118)
第二节　三硫醚类香料 ……………………………………………… (120)
第三节　四硫醚类香料 ……………………………………………… (122)
参考文献 …………………………………………………………………… (123)

第七章　一硫代缩羰基类香料 ……………………………………………… (125)
第一节　一硫代缩醛的一般合成方法 ……………………………… (125)
第二节　开链一硫代缩醛类香料 …………………………………… (129)
第三节　环状一硫代缩醛类香料 …………………………………… (131)
第四节　环状一硫代缩酮类香料 …………………………………… (132)
参考文献 …………………………………………………………………… (133)

第八章　二硫代缩羰基类香料 ……………………………………………… (138)
第一节　二硫代缩羰基化合物的一般合成方法 …………………… (138)
第二节　二硫代缩醛类香料 ………………………………………… (140)

第三节　1,3-二硫杂环戊烷类香料 …………………………………… (142)
　　第四节　其他二硫代缩酮类香料 ……………………………………… (146)
　　参考文献 ……………………………………………………………………… (147)
第九章　硫代羧酸酯类香料 ………………………………………………… (154)
　　第一节　硫代羧酸酯类香料的一般合成方法 ………………………… (154)
　　第二节　硫代羧酸酯类香料 …………………………………………… (156)
　　参考文献 ……………………………………………………………………… (159)
第十章　噻吩类香料 ………………………………………………………… (161)
　　第一节　噻吩类化合物的一般合成方法 ……………………………… (161)
　　第二节　噻酚类香料 …………………………………………………… (164)
　　参考文献 ……………………………………………………………………… (169)
第十一章　噻唑类香料 ……………………………………………………… (172)
　　第一节　噻唑类化合物的一般合成方法 ……………………………… (173)
　　第二节　烷基噻唑类香料 ……………………………………………… (176)
　　第三节　烷氧基噻唑类香料 …………………………………………… (178)
　　第四节　酰基噻唑类香料 ……………………………………………… (179)
　　第五节　二氢噻唑类香料 ……………………………………………… (181)
　　第六节　其他噻唑类香料 ……………………………………………… (182)
　　参考文献 ……………………………………………………………………… (184)
第十二章　其他含硫香料 …………………………………………………… (187)
　　第一节　α,α-二硫醇类香料 ……………………………………………… (187)
　　第二节　其他结构的含硫香料 ………………………………………… (187)
　　参考文献 ……………………………………………………………………… (192)

第一章 绪 论

第一节 含硫香料的历史

人类使用香料的历史已有 5000 多年,最初都是直接使用天然香料原料,而后是使用从天然香料中提取的精油、酊剂、浸膏、油树脂以及含有天然芳香成分的香脂等制品。合成香料的使用只有 160 多年的历史,1843 年 Cahours 合成了水杨酸甲酯,用以配制食用香精。此后,合成香料得到快速发展:1853 年,Cannizarro 用苯甲醛与碱反应合成了苯甲醇;1856 年,Perkin 用苯甲醛与乙酸酐反应合成了肉桂酸;1858 年,Berthelot 用莰烷氧化得到合成樟脑;1868 年,Perkin 合成了香豆素;1875 年,Tieman 与 Reimer 合成了香兰素[1],这些合成香料至今还在大量使用。

人类从开始食用野葱、野韭菜及烹熟了的肉类食品时就已经在享受含硫化合物提供的香味,但人类在漫长的历史长河中并没有认识到食品中含硫化合物的客观存在和作用。人类真正认识到含硫化合物对食品风味的重要性是最近 50 年的事。除了科技发展水平的限制之外,其中重要原因之一是含硫化合物在食品中的含量非常低,即便是在含硫化合物对香味起主导作用的食物中,其在食物中的含量也是常规化学分析方法难以检测到的。

含硫化合物对一些食物的风味具有核心影响力,这些食物包括肉类、发酵豆制品、咖啡、芝麻油、洋葱、大蒜、大葱、细香葱、韭菜、萝卜、甘蓝等。

20 世纪 60 年代以后,随着气相色谱(GC)、液相色谱(LC)、气相色谱-质谱联用仪(GC-MS)和液相色谱-质谱联用仪(LC-MS)等分析仪器在食品香成分分析方面的应用,人们从食品的香成分中发现了越来越多的含硫化合物。例如,1960 年从煮牛肉中发现的挥发性成分有 6 种,其中只有一种含硫化合物二甲基硫醚;1966 年从牛肉香成分中发现了硫化氢、甲硫醇、乙硫醇、丙硫醇、丁硫醇、二甲基硫醚、二甲基二硫醚等含硫化合物;1973 年在煮牛肉挥发性成分中发现了甲基丙基硫醚、甲基烯丙基硫醚、二烯丙基硫醚、二甲基二硫醚等含硫化合物;到 1975 年,从牛肉中发现的挥发性成分已有 139 种,其中含硫化合物有 53 种[2]。迄今为止,从牛肉中发现的挥发性成分有 1000 多种,其中含硫化合物有几百种。含硫化合物在食品中的含量非常低,一般在百万分之一甚至更低。进一步的研究发现,含硫化合物在浓度很低时的香味效果与高浓度时发生了非常大的变化,产生了令人愉快的食品的香味,它们对食品的特征香味有很大的影响。从此,含硫化合物作为一类新的合成香料开始被人们认识、接受,并受到重视。

含硫香料作为一类新的合成香料其历史是各类合成香料中最短的,只有 40 多年。但人类合成含硫化合物的时间并不比其他香料化合物晚。早在 1840 年就合成了甲硫醚、1844 年合成了烯丙基硫醚[3],绝大部分含硫香料在允许用作香料之前的许多年,其合成方法已经成熟,这也是含硫香料的一个特点。

含硫化合物纯品一般都具有强烈的令人不愉快的气味,很难把它们与香料联系起来。一些低分子硫化物如二甲基硫醚等常常被用作煤气赋臭剂来对煤气泄漏情况进行报警。低浓度的二甲基硫醚蒸气一般会引起恶心,食欲减退;高浓度的二甲基硫醚蒸气会对中枢神经系统有麻痹作用[4]。因此,二甲基硫醚一般被列为有毒物质。但微量的二甲基硫醚是允许使用的食品香料。FEMA(Flavor and Extract Manufactures Association of the United States,美国食品香料与萃取物制造者协会)号 2746,天然存在于橙汁、西红柿、干酪、日本米酒、红醋栗、洋葱、卷心菜等食物中。二甲基硫醚在 0.5% 浓度时具有奶油、扇贝、浆果、蔬菜样香味,在 1mg/kg 时具有西红柿、玉米、芦笋、奶制品香味和淡淡的薄荷尾香,可用于草莓、悬钩子、西红柿、炼乳、玉米、奶油、鱼、贝类、可可、鸡蛋等食品香精[5]。在饮料、冰淇淋、糖果、焙烤食品、果冻、布丁、果汁等食品中的用量为 0.13~1.4mg/kg[6]。其他与二甲基硫醚情况类似的硫化物还有硫化氢(FEMA 号 3779)、甲硫醇(FEMA 号 2716)等。

1965 年,FEMA 公布的 GRAS(generally recognized as safe,一般认为安全)物质共 1124 种,其中含硫香料有 12 种,标志着含硫香料正式得到认可。这 12 种含硫香料是:甲硫醇、烯丙硫醇、糠硫醇、2-巯基噻吩、苄硫醇、甲硫醚、烯丙基硫醚、烯丙基二硫醚、丁硫醚、3-甲硫基丙醛、3-甲硫基丙酸甲酯和异硫氰酸烯丙酯[6]。含硫香料在 1124 种有 FEMA 号的 GRAS 物质中占 1.1%。

最近 40 年,含硫香料的品种一直在增加,在近几年 FEMA 新公布的 GRAS 物质中,含硫香料的比例很高。例如,GRAS 20[7] 和 GRAS 21[8] 公布的食品香料中,含硫香料所占比例都超过了 31%。到 2005 年有 FEMA 号的含硫香料增加到 270 多种,占全部 2253 种有 FEMA 号的香料的 12%。含硫香料作为一类新的香料化合物在合成香料中已经占有非常重要的地位。可以预计,在未来二三十年内,含硫香料的品种还将保持比较快的增加速度。主要原因有三个:①食品香味分析技术在继续进步,不断有新的含硫化合物在食品香成分中被发现,或有含硫化合物在不同食品中的新发现;②方便面、肉制品、鸡精等食品和调味品工业的发展,促进了咸味香精工业的发展,而含硫化合物是咸味香精的关键香成分;③含硫香料研究开发速度和产业化速度的提高。

第二节　含硫香料的特点

一、含硫香料的阈值

阈值是与空白对照人类嗅觉或味觉能够辨别出香料香气或味道的最低浓度,

分为香气阈值和味道阈值。一般用 mg/kg、μg/kg、ng/kg 或 mg/L、μg/L、ng/L 表示。阈值与介质有关,在香味化学研究中,一般评价香味物质在水、空气中的香气阈值或味道阈值,有时根据需要也评价其在牛奶、咖啡、啤酒、茶水等中的阈值。

香气阈值或味道阈值是通过人的评价得出的,由于评价人的主观和客观条件的差异,不同人的评价结果是有差异的,有的甚至差别很大。

一般而言,含硫香料化合物是各类香料化合物中阈值最低的一类,其次是含氮香料化合物。迄今为止发现的香气阈值最低的香料化合物是二(2-甲基-3-呋喃基)二硫醚,在空气中的香气阈值是 $0.0007\sim 0.0028\text{ng/L}$[9]。为便于比较,将三个典型的含 O、S、N 的香料化合物在水中的香气阈值列于表 1-1 中。

表 1-1　含 O、S、N 的香料化合物在水中的香气阈值比较(单位:μg/kg)

香料名称	分子式	在水中的香气阈值	香料类型
麦芽酚	$C_6H_6O_3$	3.5×10^4	含 O 香料
2-乙酰基吡嗪	$C_6H_6N_2O$	6.2×10	含 N 香料
糠硫醇	C_5H_6OS	5.0×10^{-3}	含 S 香料

含硫香料化合物的低阈值使它们具有香势强和用量小的优点。含硫香料在香精和最终加香产品中的用量比含 O 或含 N 类香料小,表 1-2 列举了三个典型的含 O、S、N 的香料化合物在焙烤食品中的建议用量。

表 1-2　含 O、S、N 的香料化合物在焙烤食品中的建议用量比较(单位:mg/kg)

香料名称	分子式	在肉制品中的建议用量	香料类型
麦芽酚	$C_6H_6O_3$	30.0	含 O 香料
2-乙酰基吡嗪	$C_6H_6N_2O$	5.0	含 N 香料
糠硫醇	C_5H_6OS	2.1	含 S 香料

香料在各种食品中的建议用量是由有经验的调香师和食品科学家通过反复试验提出的,是由香料自身内在的因素——香势和香味特征决定的。任何一种香料在使用时都具有"自我限量"(self-limiting)特性,即用量低于一定范围时没有香味效果,使用者必须提高其用量;用量超过一定范围时其香味会令人不愉快,使用者不得不降低其用量。

含硫香料的产销量比其他合成香料如醛类、酮类、缩羰基类、酯类等要小得多。含硫香料的生产规模一般比较小。大部分含硫香料在刚开始阶段是在类似实验室用的玻璃反应器中生产的,许多含硫香料的主体反应釜只有几十升。但随着需求量的增加,含硫香料生产规模也在扩大。以 2-甲基-3-呋喃硫醇为例:1990 年,中国的生产装置为 5L 的玻璃反应器,每年的产销量只有几千克;2006 年,中国最大的生产装置为 5000L 的搪瓷反应釜,已经与大品种香料的生产装置规模相当。

含硫香料的生产和销售一般以千克为计量单位。生产批量小也是导致含硫香料价格高的原因之一。含硫香料的价格一般为每千克几百元到几万元人民币,每千克十几万元甚至高达几十万元人民币的也有。

从整体而言,含硫香料的价格是各类香料中最高的,但其价格近年也在不断降低,如二糠基二硫醚:1994年中国产的价格是6000元/kg;2005年的是500元/kg。价格降低的原因主要有三个方面:一是技术进步使产率提高、生产成本降低;二是生产规模扩大带来的规模效益;三是生产厂家增多导致市场竞争加剧。令人欣慰的是,含硫香料在价格降低的同时,产品质量在不断提高。

二、含硫香料的香味特征和用途

含硫香料的香味特征主要表现为与食物特别是与副食和菜肴有关的香味,如各种肉香、海鲜、咖啡、葱、蒜、洋葱、韭菜、甘蓝以及热带水果等的香味特征。例如,2-甲基-3-呋喃硫醇具有肉香、鱼香、烤肉、烤鸡香气与鱼、肉、大麻哈鱼、金枪鱼、烤香味道;糠硫醇具有鸡蛋、肉、咖啡、菜肴、芝麻油香气与烤香、洋葱、大蒜、咖啡、芝麻味道;2-甲硫基乙醛具有蔬菜、洋葱、大蒜、芥菜、坚果、土豆香气与葱蒜、蔬菜、洋葱、土豆味道;二丙基三硫具有洋葱、葱蒜、青香、热带水果香气和味道。

从含硫香料的香味特征即可以判断出含硫香料主要应用于食品香精,尤其是咸味香精和热带水果类香精。在日用香精中很少使用含硫香料,少有的一个例子是丁硫醚,它具有花香、青叶、洋葱、葱蒜、辣根、蔬菜香气和味道,它能赋予香精青香韵,除了用于水果、洋葱、大蒜、韭菜、芹菜、蘑菇、肉味等食用香精外,也可以用于花香和果香型日用香精。

三、天然存在的含硫香料化合物

迄今为止,在几乎所有的天然食物中都发现了含硫香料化合物的存在。例如,牛肉、鸡肉、猪肉、羊肉、海鲜、豆腐乳、酱油、豆豉、芥末、辣根、洋葱、大蒜、韭葱(leek)、土豆、西红柿、芦笋、萝卜、菜花、甘蓝、豌豆、蘑菇、西芹、花生、咖啡、可可、芝麻油、啤酒、啤酒花、牛奶、奶油、奶酪、面包、鸡蛋、白酒、葡萄酒、威士忌酒、茶叶、榛子、甜玉蜀黍、芫菁、阿魏、番木瓜、泡菜、龙眼、卷心菜、西番莲、菠萝、葡萄等[10]。食品香成分分析是一项长期、逐步深入、不断完善的研究工作,随着科学技术的进步和分析仪器的发展,从食品中发现的挥发性成分逐渐增多,含硫化合物也是如此。

在各种肉中发现的挥发性成分有1000多种,其中有一些具有肉香味或与肉近似的香味。尽管肉香味不能归因于某种单一的组分或某类特殊的化合物,但含硫化合物在肉香味中起着非常重要的作用,其中2-甲基-3-呋喃硫醇、2-甲基-3-甲硫基呋喃、甲基2-甲基-3-呋喃基二硫醚、二(2-甲基-3-呋喃基)二硫醚是公认的最重要的关键性肉香味含硫化合物,它们都是在肉中发现的挥发性香成分。表1-3列出它们在水中的香气阈值和味道阈值[11]。

表1-3 关键性肉香味含硫化合物在水中的阈值(单位:ng/kg)

名称	化学结构式	香气阈值	味道阈值
2-甲基-3-呋喃硫醇		5~10	2~25
2-甲基-3-甲硫基呋喃		50	5
甲基 2-甲基-3-呋喃基二硫醚		10	0.1~1
二(2-甲基-3-呋喃基)二硫醚		0.02	2

在煮牛肉中发现的重要含硫香料化合物有甲硫醇、丁硫醇、异丁硫醇、叔丁硫醇、叔戊硫醇、己硫醇、庚硫醇、仲辛硫醇、糠硫醇、2-甲基-3-呋喃硫醇、2-甲基-3-甲硫基呋喃、二(2-甲基-3-呋喃基)二硫醚、苯硫酚、2-甲基苯硫酚、2-叔丁基苯硫酚、2,6-二甲基苯硫酚、1,2-乙二硫醇、1,3-丙二硫醇、1,4-丁二硫醇、1,5-戊二硫醇、1,6-己二硫醇、甲基丙基硫醚、烯丙基甲基硫醚、二乙基硫醚、甲基丁基硫醚、乙基丁基硫醚、乙基异丁基硫醚、二丁基硫醚、二戊基硫醚、二异戊基硫醚、二烯丙基硫醚、乙烯基苯基硫醚、二丙烯基硫醚、二甲基二硫醚、二乙基二硫醚、二异丙基二硫醚、二丁基二硫醚、二异丁基二硫醚、二叔丁基二硫醚、二戊基二硫醚、二甲基三硫醚、噻吩、2-甲基噻吩、2-叔丁基噻吩、3-叔丁基噻吩、2-甲基四氢噻吩、2,5-二甲基四氢噻吩、2-噻吩醛、1,3-二硫杂环戊烷、2-甲基-1,3-二硫杂环戊烷、3,5-二甲基-1,2,4-三硫杂环戊烷(异构体混合物)、1,3-二噻烷、1,4-二噻烷、5-甲硫基糠醛、2-乙酰基噻唑、苄硫醇、2-乙酰基噻吩、5-甲基-2-甲酰基噻吩、2,5-二甲基-3-乙酰基噻吩等[12,13]。

在高压煮牛肉中发现的含硫香料化合物有甲硫醇、异丁硫醇、萘硫醇、二甲基硫醚、二甲基二硫醚、二乙基二硫醚、甲基乙基二硫醚、甲基乙烯基二硫醚、噻吩、2-甲基噻吩、2-乙基噻吩、2-丁基噻吩、2-叔丁基噻吩、3-叔丁基噻吩、2-戊基噻吩、辛基噻吩、十四烷基噻吩、2-乙酰基噻吩、3-乙酰基噻吩、5-甲基-2-乙酰基噻吩、2-噻吩基丙烯醛、2-噻吩醛、5-甲基-2-噻吩醛、2,5-二甲基-3-噻吩醛、2-丙酰基噻吩、2-甲基-5-丙酰基噻吩、2-噻吩甲硫醇、四氢噻吩-3-酮、2-甲基四氢噻吩-3-酮、噻唑、2-甲基噻唑、4-甲基噻唑、2,4-二甲基噻唑、5-乙基-4-甲基噻唑、4-乙基-2-甲基噻唑、2,4,5-三甲基噻唑、2,4-二甲基-5-乙烯基噻唑、2-乙酰基噻唑、苯并噻唑、3,5-二甲基-1,2,4-三硫杂环戊烷(异构体混合物)、5,6-二氢-2,4,6-三甲基-1,3,5-二噻嗪、2,4,6-三甲基三噻烷、2,2,4,4,6,6-六甲基三噻烷、硫代乙酸甲酯、二甲硫醇缩乙醛等。

在牛肉罐头中发现的含硫香料化合物有甲硫醇、硫化氢、甲基乙基硫醚、二甲基硫醚、环硫乙烷、硫杂环丁烷、氧硫化碳、二硫化碳、二甲基二硫醚、二甲基三硫、噻吩、2-甲基噻吩、3-甲基噻吩、2-戊基噻吩、2,3-二甲基噻吩、2,5-二甲基噻吩、2-甲酰基噻吩、3,5-二甲基-1,2,4-三硫杂环戊烷(异构体混合物)、二(甲硫基)甲烷等[13]。

在猪肉中发现的含硫香料化合物有 1-戊硫醇、2-甲基-1-丁硫醇、3-甲基-1-丁硫醇、苄硫醇、糠硫醇、2-甲基-3-呋喃硫醇、甲基 2-甲基-3-呋喃基二硫醚、二(2-甲基-3-呋喃基)二硫醚、乙基异丙基二硫醚、二丙基二硫醚、二异丙基二硫醚、甲基异丁基二硫醚、甲基 2-甲基丁基二硫醚、甲基丙基三硫醚、二甲基四硫醚、4-甲基-2,3,5-三硫杂己烷、2-甲基-4,5-二氢噻吩、2-甲基噻吩、2,5-二甲基噻吩、2-乙基噻吩、3-乙基噻吩、2-正丙基噻吩、3-正丙基噻吩、2-正丁基噻吩、3-正丁基噻吩、3,4-二乙基噻吩、2-正戊基噻吩、2-正己基噻吩、2-正庚基噻吩、3-正庚基噻吩、2-正辛基噻吩、四氢噻吩-3-酮、2-甲基四氢噻吩-3-酮、5-甲基四氢噻吩-3-酮、2,5-二甲基四氢噻吩-3-酮、2-(1-羟基乙基)噻吩、5-甲基-4,5-二氢-2-噻吩甲醛、苯并噻吩、2-甲基噻吩并[3,2-b]噻吩、3-甲基噻吩并[3,2-b]噻吩、咖啡呋喃(kahweofuran)、3-甲基-1,2,4-三噻烷、2,4,5-三甲基噻唑、2,5-二甲基-4-乙基噻唑、4,5-二甲基-2-异丙基噻唑、2-甲基-3-噻唑啉、4,5-二甲基-3-噻唑啉、2,4,5-三甲基-3-噻唑啉等[11]。

在炸鸡挥发性香成分中发现的含硫香料化合物有 2-甲基噻吩、2-异丙基噻吩、2-丁基噻吩、2-戊基噻吩、2-乙酰基噻吩、噻唑、2-甲基噻唑、2-甲基-4-乙基噻唑、2-甲基-5-乙基噻唑、2,4,5-三甲基噻唑、2,4-二甲基-5-乙基噻唑、2-异丙基-4,5-二甲基噻唑、2,5-二甲基-4-丁基噻唑、2-异丙基-4-乙基-5-甲基噻唑、2-丁基-4,5-二甲基噻唑、2-丁基-4-甲基-5-乙基噻唑、2-戊基-4,5-二甲基噻唑、2-己基-4,5-二甲基噻唑、2-庚基-4,5-二甲基噻唑、2-庚基-4-乙基-5-甲基噻唑、2-辛基-4,5-二甲基噻唑、2,4-二甲基-3-噻唑啉、2,4,5-三甲基-3-噻唑啉、3,5-二甲基-1,2,4-三噻烷、2,4,6-三甲基-1,3,5-三噻烷等[14]。

在高压炖母鸡挥发性香成分中发现的含硫香料化合物有 2-甲基-3-呋喃硫醇、3-甲硫基丙醛、糠硫醇、2,4,6-三甲基四氢-1,3,5-噻二嗪、3,5-二甲基-1,2,4-三硫杂环戊烷、5,6-二氢-2,4,6-三甲基-4H-1,3,5-二噻嗪、5,6-二氢-2,4,6-三甲基-4H-1,3,5-噻二嗪等[15]。

在鸡汤中发现的重要含硫香料化合物有 2-甲基噻吩、2-甲基-3-呋喃硫醇、2,5-二甲基-3-呋喃硫醇、糠硫醇、3-巯基-2-戊酮、甲碘磺酸钠、2,4,5-三甲基噻唑、2-甲酰基噻吩、2-甲酰基-5-甲基噻吩、2-乙酰基噻吩、2-乙酰基噻唑、2-乙酰基-2-噻唑啉等[9]。

在炒榛子挥发性香成分中发现的含硫香料化合物有甲硫醇、二甲基硫醚、二甲基二硫醚、二乙基二硫醚、二甲基三硫醚、3-甲硫基丙醛、二氢噻吩-3-酮、2-噻吩甲醛、4-甲基-5-乙烯基噻唑、苯并噻唑、3,5-二甲基-1,2,4-三硫杂环戊烷等[16]。

在牛奶中发现的挥发性化合物有 400 多种,其中对牛奶香气有贡献的含硫香料化合物有硫化氢、甲硫醇、异丁硫醇、二甲基硫醚、二甲基二硫醚、3-甲硫基丙醛、二甲基砜、硫氰酸苄酯、异硫氰酸甲酯、异硫氰酸乙酯、异硫氰酸 3-丁烯醇酯等[10,17]。

在大蒜中发现的含硫化合物有甲硫醇、二甲基硫醚、甲基丙基硫醚、甲基烯丙基硫醚、二丙基硫醚、二烯丙基硫醚、二甲基二硫醚、甲基丙基二硫醚、甲基烯丙基二硫醚、烯丙基丙基二硫醚、二丙基二硫醚、二烯丙基二硫醚、二甲基三硫醚、甲基丙基三硫醚、甲基烯丙基三硫醚等[10]。

在洋葱中发现的含硫化合物有甲硫醇、乙硫醇、丙硫醇、烯丙硫醇、2-羟基丙硫醇、二甲基硫醚、甲基烯丙基硫醚、甲基丙烯基硫醚、二丙烯基硫醚、二烯丙基硫醚、丙基烯丙基硫醚、丙基丙烯基硫醚、二甲基二硫醚、甲基丙基二硫醚、甲基烯丙基二硫醚、甲基烯丙基二硫醚、二丙基二硫醚、丙基异丙基二硫醚、丙基丙烯基二硫醚、丙烯基烯丙基二硫醚、丙烯基烯丙基二硫醚、二烯丙基二硫醚、二甲基三硫醚、甲基丙基三硫醚、甲基丙烯基三硫醚、甲基烯丙基三硫醚、二丙基三硫醚、丙基异丙基三硫醚、丙基丙烯基三硫醚、丙基烯丙基三硫醚、二烯丙基三硫醚、二异丙基三硫醚、二甲基四硫醚、2,4-二甲基噻吩、2,5-二甲基噻吩、3,4-二甲基噻吩、3,4-二甲基-2,5-二氢噻吩-2-酮等[10]。

在韭葱中发现的含硫化合物有甲硫醇、乙硫醇、丙硫醇、甲基烯丙基硫醚、丙基烯丙基硫醚、二烯丙基硫醚、二甲基二硫醚、甲基丙基二硫醚、丙基丙烯基二硫醚、甲基烯丙基二硫醚、二烯丙基二硫醚、丙烯基烯丙基二硫醚、二甲基三硫醚、甲基丙基三硫醚、丙基丙烯基三硫醚、3,4-二甲基-2,5-二氢噻吩-2-酮、苯并噻唑、硫代亚磺酸二甲酯、硫代亚磺酸甲丙酯、硫代亚磺酸甲烯丙酯、硫代亚磺酸二丙酯、硫代亚磺酸丙丙烯酯、硫代亚磺酸烯丙丙烯酯、硫化氢、二硫化碳等[10]。

在咖啡中发现的含硫化合物有硫化氢、甲硫醇、乙硫醇、丙硫醇、糠硫醇、苄硫醇、甲硫醚、甲基乙基硫醚、甲基糠基硫醚、甲基 5-甲基糠基硫醚、甲苯基硫醚、甲基 2-羟基苯基硫醚、二糠基硫醚、1-甲硫基-2-丁酮、二甲基二硫醚、甲基乙基二硫醚、二乙基二硫醚、硫代糠酸甲酯、硫代乙酸糠酯、硫代丙酸糠酯、噻吩、2-甲基噻吩、3-乙烯基噻吩、2-甲基-4-乙基噻吩、2-丙基噻吩、2-丁基噻吩、2-噻吩基甲醇、2-甲酰基噻吩、2-甲酰基-3-甲基噻吩、2-甲酰基-5-甲基噻吩、2-乙酰基噻吩、2-乙酰基-3-甲基噻吩、2-乙酰基-4-甲基噻吩、2-乙酰基-5-甲基噻吩、2-苯并噻吩、四氢噻吩-3-酮、2-甲基四氢噻吩-3-酮、噻唑、2-甲基噻唑、4-甲基噻唑、5-甲基噻唑、2,4-二甲基噻唑、2,5-二甲基噻唑、4,5-二甲基噻唑、2,4-二甲基-5-乙基噻唑、2,5-二甲基-4-乙基噻唑、4,5-二甲基-2-乙基噻唑、三甲基噻唑、5-乙基噻唑、2-乙基-

4-甲基噻唑、4-乙基-2-甲基噻唑、5-乙基-2-甲基噻唑、5-乙基-4-甲基噻唑、4-乙基-5-甲基噻唑、2,4-二乙基噻唑、2,5-二乙基噻唑、2-丙基-4-甲基噻唑、4-丁基噻唑、苯并噻唑、2-乙酰基-4-甲基噻唑等[10]。

表 1-4 对在各种肉挥发性成分中发现的部分含硫化合物进行了归纳,这些发现对于肉香味化学研究有重要的参考价值。

表 1-4　在肉类挥发性成分中发现的部分含硫化合物

化合物名称	天然存在	参考文献
1-(2-甲基-3-呋喃硫基)乙硫醇	猪肉	[11]
1,1-二甲硫基乙烷	牛肉	[10]
1,1-乙二硫醇	猪肉	[11]
1,2-乙二硫醇	牛肉、鸡肉	[10]
1,3-丙二硫醇	牛肉	[10]
1,3-二硫杂环戊烷	牛肉	[10]
1,3-二噻烷	牛肉	[10]
1,4-丁二硫醇	牛肉	[10]
1,4-二噻烷	牛肉	[10]
1,5-戊二硫醇	牛肉	[10]
1,6-己二硫醇	牛肉	[10]
1-丙硫醇	鸡肉、牡蛎	[10]
1-丁硫醇	鸡肉、牛肉、牡蛎	[10,11]
1-庚硫醇	鸡肉、牛肉	[10,11]
1-甲硫基-1-乙硫醇	猪肉、鸡肉、牛肉	[11,18,19]
1-甲硫基-3-戊酮	鸡肉	[11]
1-戊硫醇	猪肉、鸡肉	[11]
1-辛硫醇	鸡肉	[11]
2-(1-丙烯硫基)噻吩	香螺、虾	[20]
2,2,4,4,6,6-六甲基三噻烷	牛肉	[10]
2,3-二甲基噻吩	牛肉	[10]
2,4,5-三甲基-3-噻唑啉	牛肉、猪肉、炸鸡肉	[11,18]
2,4,5-三甲基噻唑	鸡肉、香螺、虾、猪肉、鳕鱼、牛肉	[10,11,20~22]
2,4,6-三甲基三噻烷	牛肉	[10]
2,4-二甲基-5-乙基噻唑	烤牛肉、烤猪肉、炸鸡肉	[18,21]
2,4-二甲基-5-乙烯基噻唑	牛肉	[10]
2,4-二甲基噻唑	牛肉、鳕鱼	[10,21]
2,4-二甲基-3-噻唑啉	猪肉	[11,21]
2,5-二甲基-1,3,4-三硫杂环戊烷	牛肉	[2]
2,5-二甲基-3-呋喃硫醇	鸡肉、牛肉	[21,22]

续表

化合物名称	天然存在	参考文献
2,5-二甲基-3-噻吩甲醛	牛肉	[10]
2,5-二甲基-4-乙基噻唑	猪肉	[11]
2,5-二甲基噻吩	猪肉、牛肉、鳕鱼	[10,11]
2,5-二甲基四氢噻吩	牛肉	[10]
2,5-二甲基四氢噻吩-3-酮	猪肉	[11]
2,5-甲基-3-乙酰基噻吩	鸡肉	[11]
2,6-二甲基-4-丁基-1,3,5-二氢二噻嗪	对虾	[20]
2,6-二甲硫基苯酚	牛肉	[10]
2-丙基-5-异戊基噻吩	鸡肉	[10]
2-丙硫醇	牡蛎	[10]
2-丙酰基噻吩	牛肉	[10]
2-丁基-4,5-二甲噻唑	牛肉、鸡肉、腌熏肉	[22]
2-丁基-4-甲基-5-乙基噻唑	牛肉、鸡肉	[22]
2-丁基噻吩	牛肉、鸡肉、猪肉	[10,11,22,23]
2-丁硫醇	牡蛎	[10]
2-丁酰基噻吩	牛肉	[22]
2-庚基-4,5-二甲噻唑	鸡肉	[22]
2-庚基-4-甲基-5-乙基噻唑	鸡肉	[22]
2-庚基噻吩	牛肉、猪肉	[11,22]
2-庚酰基噻吩	牛肉	[22]
2-己基-4,5-二甲噻唑	鸡肉	[22]
2-己基噻吩	牛肉、猪肉	[11,22]
2-甲基-1,3-二硫杂环戊烷	牛肉	[10]
2-甲基-1-丙硫醇	牡蛎	[10]
2-甲基-1-丁硫醇	猪肉	[11]
2-甲基-2-丙硫醇	牡蛎	[10]
2-甲基-3-呋喃基 2-甲基-3-噻吩基二硫醚	牛肉	[19]
2-甲基-3-呋喃硫醇	金枪鱼、牛肉、鸡肉、猪肉	[11,18,21,23]
2-甲基-3-甲硫基呋喃	牛肉	[11,18,19,23]
2-甲基-3-噻吩硫醇	猪肉	[11]
2-甲基-3-噻唑啉	猪肉	[11]
2-甲基-3-乙硫基呋喃	鸡肉	[11]
2-甲基-4,5-二氢噻吩	猪肉	[11]
2-甲基-4-乙基噻唑	牛肉	[10]
2-甲基-5-丙酰基噻吩	牛肉	[10]

续表

化合物名称	天然存在	参考文献
2-甲基苯硫酚	牛肉	[10]
2-甲基噻吩	牛肉、鸡肉	[10]
2-甲基噻吩并[3,2-b]噻吩	猪肉	[11]
2-甲基噻唑	牛肉	[10]
2-甲基四氢噻吩	牛肉	[10]
2-甲基四氢噻吩-3-酮	猪肉、鸡肉、牛肉	[10,11]
2-甲硫基甲基-2-丁烯醛	猪肉、鸡肉、牛肉	[11]
2-甲酰基-5-甲基噻吩	牛肉、鸡肉	[22]
2-甲酰基噻吩	牛肉	[10]
2-噻吩丙烯醛	牛肉	[10]
2-噻吩甲醇	牛肉	[10]
2-噻吩甲醛	鸡肉	[11]
2-叔丁基噻吩	牛肉	[10]
2-叔丁硫基苯酚	牛肉	[10]
2-戊基-4,5-二甲基噻唑	牛肉、鸡肉	[22]
2-戊基噻吩	牛肉	[10]
2-戊基噻唑	牛肉、鸡肉、猪肉	[11,22]
2-辛基-4,5-二甲基噻唑	鸡肉	[22]
2-辛基噻吩	牛肉、猪肉	[11,22]
2-辛酰基噻吩	牛肉	[22]
2-乙基-4,5-二甲基噻唑	鳕鱼	[10]
2-乙基噻吩	猪肉、牛肉	[10,11]
2-乙酰基-5-甲基噻吩	鸡肉	[11]
2-乙酰基噻吩	牛肉、鸡肉	[10,18]
2-乙酰基噻唑	烤猪肉、羊肉、猪肝、牛肉、肉、鳕鱼、对虾	[10,11,18,20,21]
2-正丙基噻吩	猪肉、鸡肉	[11]
3,4-二乙基噻吩	猪肉	[11]
3,5,6-三甲基-1,2,4-二噻嗪	对虾	[20]
3,5-二甲基-1,2,4-三硫杂环戊烷	鸡肉、香螺、虾、牛肉、对虾	[10,20,24]
3,6-二甲基-1,2,4,5-四噻烷	猪肉、鸡肉	[11]
3-丁基噻吩	猪肉	[11]
3-庚基噻吩	猪肉	[11]
3-甲基-1,2,4-四噻烷	猪肉	[11]
3-甲基-1-丁硫醇	猪肉、牡蛎	[10,11]
3-甲基-2-噻吩甲醛	鸡肉	[11]

续表

化合物名称	天然存在	参考文献
3-甲基-5-丁基-1,2,4-三硫杂环戊烷	鸡肉	[22]
3-甲基-5-戊基-1,2,4-三硫杂环戊烷	鸡肉	[22]
3-甲基噻吩	猪肉、牛肉	[10,11]
3-甲基噻吩并[3,2-b]噻吩	猪肉	[11]
3-甲硫基丙醛	牛肉、鸡肉、鳗鱼	[2,20,22,25]
3-巯基-2-戊酮	牛肉、鸡肉	[22,23]
3-噻吩甲醛	鸡肉	[11]
3-噻吩硫醇	猪肉	[11]
3-叔丁基噻吩	牛肉	[10]
3-乙基噻吩	猪肉	[11]
3-乙酰基-5-甲基噻吩	鸡肉	[11]
3-乙酰基噻吩	牛肉	[10]
3-正丙基噻吩	猪肉	[11]
4,5-二甲基-2-异丙基噻唑	猪肉	[11]
4-甲基-2,3,5-三硫杂己烷	猪肉	[11]
4-甲基-2-噻吩甲醛	鸡肉	[11]
4-甲基-5-乙基噻唑	牛肉	[10]
4-甲基噻唑	牛肉	[10]
5-甲基-2-噻吩甲醛	牛肉、鸡肉	[10,11]
5-甲基-2-乙酰基噻吩	牛肉	[10]
5-甲基-4,5-二氢噻吩甲醛	猪肉	[11]
5-甲基四氢噻吩-3-酮	猪肉、鸡肉	[11]
5-甲硫基糠醛	牛肉	[2,10]
5-乙基-2,4-二甲基噻唑	鳕鱼	[10]
苯并噻吩	猪肉	[11]
苯并噻唑	鸡肉、牛肉、对虾	[10,11,20,24]
苯基乙烯基硫醚	牛肉	[10]
苯硫酚	牛肉	[10]
苄硫醇	猪肉、牛肉	[10,11]
丙基噻吩	鳕鱼	[10]
丙硫醇	牛肉	[2]
丁硫醇	牛肉	[2]
二(2-甲基-3-呋喃基)二硫醚	牛肉、猪肉	[11,18,19,21]
二丙基硫醚	鸡肉	[10]
二丙烯基硫醚	牛肉	[10]
二丁基二硫醚	牛肉	[10]

续表

化合物名称	天然存在	参考文献
二丁基硫醚	牛肉	[10]
二甲基二硫醚	牛肉、猪肝、鸡肉、牡蛎、鳕鱼、对虾	[2,10,20,24]
二甲基硫醚	牛肉、猪肝、鸡肉、牡蛎、鳕鱼、鲲鱼	[2,10,20,24]
二甲基三硫醚	牛肉、香螺、虾、鲐鱼	[10,20,24]
二甲基四硫醚	猪肉	[11]
二甲硫基甲烷	牛肉	[10]
二甲基亚砜	猪肝	[10]
二糠基二硫醚	牛肉	[11]
二硫化碳	牛肉、鸡肉	[10]
二叔丁基二硫醚	牛肉	[10]
二戊基二硫醚	牛肉	[10]
二戊基硫醚	牛肉	[10]
二烯丙基硫醚	牛肉	[2,10]
二乙基二硫醚	牛肉、鸡肉	[10]
二乙基硫醚	牛肉、鸡肉、牡蛎	[10]
二异丙基二硫醚	猪肉、牛肉	[11,10]
二异丁基二硫醚	牛肉	[10]
二异戊基硫醚	牛肉	[10]
二正丙基二硫醚	猪肉	[11]
己硫醇	牛肉、鸡肉	[10]
甲基烯丙基硫醚	牛肉	[2]
甲基 2-甲基-3-呋喃基二硫醚	牛肉、猪肉	[11,23]
甲基 2-甲基丁基二硫醚	猪肉	[11]
甲基苄基硫醚	猪肝	[10]
甲基丙基硫醚	牛肉、鸡肉、牡蛎	[2,10]
甲基丙基三硫醚	猪肉	[11]
甲基丁基硫醚	牡蛎、牛肉	[10]
甲基砜	猪肝	[10]
甲基糠基二硫醚	猪肝、高压煮猪肝	[10,18]
甲基糠基硫醚	猪肝、高压煮猪肝	[10,18]
甲基烯丙基硫醚	牛肉	[10]
甲基乙基二硫醚	牛肉	[10]
甲基乙基硫醚	牛肉、鸡肉、牡蛎	[10]
甲基乙烯基二硫醚	牛肉	[10]
甲基异丙基硫醚	鸡肉	[10]
甲基异丁基二硫醚	猪肉	[11]

续表

化合物名称	天然存在	参考文献
甲硫醇	猪肝、牛肉、鸡肉、牡蛎、鳕鱼	[2,10]
咖啡呋喃	猪肉	[11]
糠基 2-甲基-3-呋喃基二硫醚	牛肉	[19]
糠硫醇	煮牛肉、烤猪肉	[11,18,22]
糠醛二甲硫醇缩醛	鸡肉	[11]
硫代丙酸甲酯	猪肝	[10]
硫代丙酸乙酯	猪肝	[10]
硫代乙酸甲酯	猪肝、牛肉	[10]
硫化氢	牛肉、鸡肉、牡蛎、鳕鱼	[2,10,24]
硫杂环丁烷	牛肉	[10]
萘硫酚	牛肉	[10]
噻啶	牛肉	[25]
噻吩	牛肉	[10]
2-噻吩甲醛	牛肉	[10]
3-噻吩甲醛	牛肉	[10]
噻唑	牛肉	[10]
叔丁硫醇	牛肉	[10]
叔戊硫醇	牛肉	[10]
四癸基噻吩	牛肉	[10,22]
四氢噻吩-3-酮	猪肉、牛肉	[10,11]
四噻烷	鸡肉	[10]
辛基噻吩	牛肉	[10]
氧硫化碳	牛肉、鸡肉	[10]
乙基丙基硫醚	鸡肉	[10]
乙基丁基硫醚	牛肉	[10]
乙基异丙基二硫醚	猪肉	[11]
乙基异丁基硫醚	牛肉	[10]
乙硫醇	鸡肉、牡蛎、牛肉	[2,10,24]
乙烯基硫醚	牛肉	[10]
异丙基噻吩	对虾	[20]
异丁硫醇	牛肉	[10]
正十六硫醇	猪肉	[11]
仲辛硫醇	牛肉	[10]

四、在热降解和热反应产物中发现的含硫化合物

为了探索肉香味含硫化合物的生成途径,人们对酵母提取物、维生素 B_1 热降解反应、氨基酸与还原糖的热反应等进行了深入的研究,从中发现了许多重要含硫香味化合物,并对它们的形成机理进行了描述。这些研究对于揭示肉香味含硫化合物的成因和肉味香精的制备都有重要意义。

在酵母提取物挥发性成分中发现了甲硫醇、1,1-乙二硫醇、糠硫醇、5-甲基糠硫醇、2-甲基-3-呋喃硫醇、1-(2-甲基-3-呋喃硫基)乙硫醇、2-甲基-3-噻吩硫醇、2-巯基-3-丁酮、1-甲硫基-1-乙硫醇、1-甲硫基-1-丙硫醇、1-甲硫基-2-甲基-1-丙硫醇、1-甲硫基-3-甲基-1-丁硫醇、二甲基二硫醚、甲基乙基二硫醚、2-甲基-3-甲硫基呋喃、二甲基三硫醚、二甲基四硫醚、二糠基硫醚、3-甲硫基丙醛、1,1-二甲硫基乙烷、2-甲硫基甲基-2-丁烯醛、3-甲基-2,4,5-三硫杂己烷、甲基 2-甲基-3-呋喃基二硫醚、二(2-甲基-3-呋喃基)二硫醚、2-甲基-3-呋喃基-2-甲基-3-噻吩基二硫醚、二[1-(2-甲基-3-呋喃硫基)乙基]二硫醚、咖啡呋喃、2-甲基噻吩、2-噻吩甲醛、3-噻吩甲醛、5-甲基-2-噻吩甲醛、5-甲基-4,5-二氢-2-噻吩甲醛、2-乙酰基噻吩、2-丙酰基噻吩、2-戊酰基噻吩、3-苯基噻吩、四氢噻吩-3-酮、3-甲基-1,2-二噻烷-4-酮、2,4,6-三甲基-1,3,5-三噻烷、3,6-二甲基-1,2,4,5-四噻烷、3,5-二甲基-1,2,4-三硫杂环戊烷、3-异丙基-5-甲基-1,2,4-三硫杂环戊烷、2,4,6-三甲基-1,3,5-二氢二噻嗪、2,6-二甲基-4-乙基-1,3,5-二氢二噻嗪、4,6-二甲基-2-乙基-1,3,5-二氢二噻嗪、2,6-二甲基-4-正丙基-1,3,5-二氢二噻嗪、4,6-二甲基-2-正丙基-1,3,5-二氢二噻嗪、2,6-二甲基-4-异丙基-1,3,5-二氢二噻嗪、4,6-二甲基-2-异丙基-1,3,5-二氢二噻嗪、2,6-二甲基-4-异丁基-1,3,5-二氢二噻嗪、4,6-二甲基-2-异丁基-1,3,5-二氢二噻嗪、4,6-二甲基-2-仲丁基-1,3,5-二氢二噻嗪、2-乙基-4-异丁基-6-甲基-1,3,5-二氢二噻嗪、4-乙基-2-异丁基-6-甲基-1,3,5-二氢二噻嗪、4-异丁基-2-异丙基-6-甲基-1,3,5-二氢二噻嗪、2-异丁基-4-异丙基-6-甲基-1,3,5-二氢二噻嗪、4-异丁基-6-异丙基-2-甲基-1,3,5-二氢二噻嗪、2,4-二异丁基-6-甲基-1,3,5-二氢二噻嗪等含硫化合物[26,27]。

维生素 B_1 与水或丙二醇在 135℃ 下搅拌加热 30min,在其挥发性产物中发现了 2-甲基-3-四氢呋喃硫醇、2-甲基-3-呋喃硫醇、2-甲基-4,5-二氢-3-呋喃硫醇、2,5-二甲基-3-呋喃硫醇、二(2-甲基-3-呋喃基)二硫醚、2-甲基-3-呋喃基 2-甲基-4,5-二氢-3-呋喃基二硫醚、二(2-甲基-4,5-二氢-3-呋喃基)二硫醚、2-甲基-4,5-二氢-3-呋喃基 2-甲基-3-四氢呋喃基二硫醚、2,5-二甲基-4,5-二氢-3-呋喃基 2-甲基-4,5-二氢-3-呋喃基三硫醚、2-甲基噻吩、2-甲基-4,5-二氢噻吩、2-甲基四氢噻吩、2-甲基四氢噻吩-3-酮、2-乙酰基噻吩、4-甲基噻唑、4,5-二甲基噻唑、5-乙基-4-甲基噻唑、4-甲基-5-羟乙基噻唑、4-甲基-5-(2-氯乙基)噻唑、二甲基二硫醚、硫代乙酸甲酯、3-巯基-2-戊酮、3-甲基-4-氧代-1,2-二噻烷、1-甲基二环[3.3.0]-2,8-二噁-4-硫杂辛烷、

1,3-二甲基二环[3.3.0]-2,8-二噁-4-硫杂辛烷等含硫化合物[28]。

维生素 B_1 与丙二醇在微波中加热 5 min,在其挥发性成分中发现了 2-甲基-3-呋喃硫醇、5-羟甲基异噻唑、4,5-二甲基噻唑、硫代乙酸、二乙酰基硫醚、硫氰酸丙酯、1-甲硫基乙硫醇乙酸酯、2-乙酰基噻吩、2-叔丁硫基乙酸、2-乙酰基-3-甲基噻吩、硫代苯甲酸、2-巯基-3-羟基吡啶、2-乙基-5-丙基噻吩、3-(2-丁烯基)噻吩、2,5-二甲基-3-乙酰基噻吩、2-甲酰基-2,3-二氢噻吩、2-丙基噻烷、1-甲硫基吩嗪、2,3,4,5-四硫二环[4.4.0]癸烷、乙酸 1-巯基-2-丙醇酯、2-氨基-2-噻唑啉-4-羧酸、2-噻吩基二硫醚、3-(4-甲基-5-顺-苯基-1,3-烷-2-)噻酚、2-[3-(2-噻吩基)苯基]噻吩、2-甲硫基吩嗪、三甲硫基乙烯、2-丁硫基四氢-$2H$-吡喃、二叔十二烷基二硫醚等含硫化合物[29]。

胱氨酸、维生素 B_1、维生素 C、谷氨酸单钠盐和水组成的模拟肉香味体系在 pH 为 5.0、120℃下加热 30min 产生的挥发性成分中发现了 70 种含硫化合物,包括 2-甲基-3-呋喃硫醇、2-甲基-3-噻吩硫醇、二(2-甲基-3-呋喃基)二硫醚、二(2-甲基-3-噻吩基)二硫醚、2-甲基-3-呋喃基 2-甲基-4,5-二氢-3-呋喃基二硫醚、二(2-甲基-4,5-二氢-3-呋喃基)二硫醚、2-甲基-3-呋喃基 2-甲基-3-噻吩基二硫醚、2-甲基-3-(顺-2-甲基-3-四氢噻吩硫基)呋喃、2-甲基-3-(反-2-甲基-3-四氢噻吩硫基)呋喃、2-甲基-3-(顺-2-甲基-3-四氢噻吩硫基)噻吩、2-甲基-3-(反-2-甲基-3-四氢噻吩硫基)噻吩、2-甲基-3-(2-四氢噻吩甲硫基)呋喃、2-甲基-3-(2-四氢噻吩甲硫基)噻吩、2-甲基-2-(2-甲基-3-噻吩基)四氢噻吩、2-甲基-2-(2-甲基-3-呋喃基)四氢噻吩等[30]。

在牛肉蛋白酶解物、L-半胱氨酸盐酸盐、D-木糖和 D-葡萄糖的热反应产物中发现了糠硫醇、2-甲基-4,5-二氢-3-呋喃硫醇、2-甲基-3-呋喃硫醇、甲基 2-甲基-3-呋喃基二硫醚、二(2-甲基-3-呋喃基)二硫醚、二糠基二硫醚、二乙基二硫醚、3-巯基-2-丁硫醇、二乙烯基硫醚、3-巯基-2-戊酮、四氢-3a-甲基-1,3-二杂环戊烯基并[4,5-b]呋喃、噻唑、4-甲基噻唑、4-甲基-5-羟乙基噻唑、4-甲基-5-乙烯基噻唑、4-甲基-5-(2-氯乙基)噻唑、2-乙酰基噻唑、4-甲基-5-(甲氧基羰基甲基)-1,2-异噻唑、2-甲基噻吩、3-甲基-2-噻吩甲醛、3-乙基-2-噻吩甲醛、2-丙烯硫基噻吩、2-甲基-3(2H)-二氢噻吩酮、2-甲基-4,5-二氢噻吩等[31]。

在鸡肉酶解物的热反应产物中也发现了重要肉香味含硫化合物二(2-甲基-3-呋喃基)二硫醚[32]。

第三节 中国的含硫香料工业

中国的含硫香料工业已经形成了比较完整的研发和生产体系,目前生产的含硫香料有 100 多个品种,这些含硫香料分别属于硫醇、硫醚、二硫醚、多硫醚、硫代羧酸酯、缩硫羰基、含硫杂环等化合物类型[33]。

一、硫醇类香料

甲硫醇是结构最简单的硫醇类香料,由于常温下是气体,调香中一般只使用其溶液。甲硫醇主要用于蛋氨酸、医药、农药和精细化工中间体合成,中国的年使用量约4万t,其作为香料的用量可以忽略。

糠硫醇是中国生产较早的一个硫醇类香料,在20世纪80年代就已批量生产。在中国含硫香料发展的过程中,2-甲基-3-呋喃硫醇及其衍生物是具有里程碑地位的,其生产技术的推广和应用获得了香料领域第一个国家科学技术进步二等奖,其产品已经应用于几乎所有的中、高档咸味香精,并大量出口美国、日本及欧洲各国。目前,中国生产的硫醇类香料主要有丙硫醇、糠硫醇、2-甲基-3-呋喃硫醇、2-甲基-3-四氢呋喃硫醇、2,5-二甲基-3-呋喃硫醇、3-巯基-1-丙醇、烯丙硫醇、1-丁硫醇、1,2-丙二硫醇、1,3-丙二硫醇、1,2-丁二硫醇、1,4-丁二硫醇、2,3-丁二硫醇、3-巯基-2-丁醇、3-巯基-2-丁酮、1,6-己二硫醇、1,8-辛二硫醇、1,9-壬二硫醇、苄硫醇、2-苯乙硫醇、α-甲基-β-羟基丙基 α′-甲基-β′-巯基丙基硫醚、2-巯基吡嗪、硫代乳酸、硫代乳酸乙酯、硫代薄荷酮、硫代松油醇、硫代香叶醇等。

二、硫醚类香料

二甲基硫醚是结构最简单的硫醚类香料,其在香料中的地位远不及作为有机合成中间体、有机溶剂及煤气赋臭剂等重要,其用量也相差悬殊。在硫醚类香料中需要特别指出的是2-甲基-3-甲硫基呋喃,这是于1986年从煮牛肉中发现的肉香味化合物,中国于1990年研制成功,该香料于2000年获得FEMA号[34]。目前,中国生产的硫醚类香料主要有2-甲基-3-甲硫基呋喃、甲基糠基硫醚、甲基苄基硫醚、二糠基硫醚、甲基乙基硫醚、二乙基硫醚、二烯丙基硫醚、二丁基硫醚、烯丙基甲基硫醚、糠基异丙基硫醚、3-甲硫基丙醛、3-甲硫基丙醇、3-甲硫基-1-己醇、3-甲硫基己醛、3-甲硫基丙酸甲酯、3-甲硫基丙酸乙酯、3-糠硫基丙酸乙酯、4-糠硫基-2-戊酮、2-甲基-3-甲硫基吡嗪、2-甲基-5-甲硫基吡嗪、2-甲基-6-甲硫基吡嗪、2-甲基-3-糠硫基吡嗪、2-甲基-5-糠硫基吡嗪、2-甲基-6-糠硫基吡嗪、2-甲硫基吡嗪、2-糠硫基吡嗪等。

三、二硫醚和多硫醚类香料

二甲基二硫醚是结构最简单的二硫醚类香料,但其作为溶剂及催化剂的钝化剂等的用量远大于其在香料工业中的用量。甲基2-甲基-3-呋喃基二硫醚(国外商品代号"719")是中国第一个实现工业化生产的不对称二硫醚类香料,是肉味香精的关键性香料。目前,中国生产的二硫醚类香料主要有二丙基二硫醚、二烯丙基二硫醚、二糠基二硫醚、二(2-甲基-3-呋喃基)二硫醚、甲基丙基二硫醚、甲基烯丙基二硫醚、甲基糠基二硫醚、甲基苄基二硫醚、甲基 2-甲基-3-呋喃基二硫醚、丙基烯丙基二硫醚、丙基糠基二硫醚、丙基 2-甲基-3-呋喃基二硫醚等。目前,中国生产的多硫醚类香料主要有二甲基三硫醚等。

四、硫代羧酸酯类香料

硫代羧酸酯类香料化学性质稳定,在食品香精中应用广泛。目前,中国生产的硫代羧酸酯类香料主要有硫代甲酸糠酯、硫代乙酸甲酯、硫代乙酸乙酯、硫代乙酸丙酯、硫代乙酸糠酯、2-甲基-3-呋喃硫醇乙酸酯、2,5-二甲基-3-呋喃硫醇乙酸酯、硫代丙酸糠酯、硫代丁酸甲酯、硫代糠酸甲酯等。

五、缩硫羰基类香料

缩硫羰基类香料生产工艺简便,在技术上中国能够生产的品种已经很多[35~37],由于这类香料目前允许使用的品种较少,实际生产的品种并不多,主要有甲醛二甲硫醇缩醛、苯甲醛二甲硫醇缩醛等。

六、含硫杂环类香料

含硫杂环类香料品种最多的是噻唑类香料,中国生产的主要有噻唑、2-甲基噻唑、4-甲基噻唑、2,4-二甲基噻唑、4,5-二甲基噻唑、2-乙基-4-甲基噻唑、4-甲基-5-乙烯基噻唑、4-甲基-5-(β-羟乙基)噻唑、4-甲基-5-(β-羟乙基)噻唑乙酸酯、2-异丙基-4-甲基噻唑、2,4,5-三甲基噻唑、2-异丁基噻唑、2-甲氧基噻唑、2-甲硫基噻唑、2-乙氧基噻唑、2-乙酰基噻唑、2,4-二甲基-5-乙酰基噻唑等。其他含硫杂环类香料有 2-乙酰基噻酚、四氢噻吩-3-酮、2-甲基四氢噻吩-3-酮、2-甲基四氢噻吩-3-酮等。

参 考 文 献

[1] 孙宝国,何坚. 香料化学与工艺学. 第二版. 北京:化学工业出版社,2004. 20~21
[2] Moody W G. Beef flavor—a review. Food Technology, 1983,(3):227~238
[3] 朱瑞鸿,薛群成,李忠臣. 合成食用香料手册. 北京:中国轻工业出版社,1986. 661~663
[4] 魏文德. 有机化工原料大全,第三卷. 北京:化学工业出版社,1990. 206~208
[5] Mosciano G et al. Organoleptic characteristics of flavor materials. Perfumer & Flavorist, 1998,23(1):33~36
[6] Hall R L, Oser B L. Recent progress in the consideration of flavoring ingredients under the food additives amendment, III. GRAS substances. Food Technology, 1965,19(2): 151~197
[7] Smith R L, Doull J, Feron V J et al. GRAS flavoring substances 20. Food Technology, 2001,55(12):34~55

[8] Smith R L, Cohen S M, Doull J et al. GRAS flavoring substances 21. Food technology, 2003, 57(5): 46~59

[9] Gasser U, Grosch W. Primary odorants of chicken broth, a comparative study with meat broths from cow and ox. Z Lebensm Unters Forsch, 1990, 190: 3~8

[10] Morton I D, Macleod A J. Food Flavors, Part A. Introduction. New York: Elsevier Science Publishing Company Inc., 1982. 173~248

[11] Werkhoff P, Brüning J, Emberger R et al. Flavor chemistry of meat volatiles: new results on flavor components from beef, pork, and chicken//Hopp R, Mori K. Recent Developments in Flavor and Fragrance Chemistry. New York: VCH Publisher, 1992. 183~213

[12] Gasser U, Grosch W. Identification of volatile flavor compounds with high aroma values from cooked beef. Z Lebensm Unters Forsch, 1988, 186: 489~494

[13] 孙宝国等. 食用调香术. 北京:化学工业出版社,2003. 444~479

[14] Tang J et al. Volatile compounds from fried chicken. J. Agric. Food Chem., 1983, 31(6): 1287~1290

[15] Farkas P, Sadecka J, Kovac M et al. Key odorants of pressure-cooked hen meat. Food Chemistry, 1997, 60(4): 617~620

[16] Kinlin T E, Muralidhara R, Pitter A O et al. Volatile compounds of roasted Filberts. J. Agric. Food Chem., 1972, 20(5): 1021~1028

[17] Belitz H D, Grosch W. Food chemistry. Second Edition. New York: Springer, 1999. 508~509

[18] Mottram D S. Meat//Maarse H. Volatile compounds in foods and beverages. Zeist, The Netherlands: TNO-CIVO Food Analysis Institute, 1991. 138~159

[19] Piggott J R, Paterson A. Understanding natural flavor. New York: Blackie Academic & Professional, 1994. 154~159

[20] Shahidi F, Cadwallader K R. Flavor and lipid chemistry of seafoods. Washington DC: American Chemical Society, 1997. 36~67

[21] Teranishi R, Flath R A, Sugisawa H. Flavor research. recent advances. New York: Marcel Dekker Inc., 1981. 223~224

[22] Shahidi F. Flavor of meat and meat products. New York: Blackie Academic & Professional, 1994. 53~226

[23] Mussinan C J. Sulfur compounds in food. Washington DC: American Chemical Society, 1994. 185~278

[24] Belitz H D. Food chemistry. Second Edition. New York: Springer, 1999. 337~339

[25] Ashurst P R. Food flavorings. Second Edition. New York: Blackie Academic & Professional, 1995. 147~148

[26] Ames J M, Leod G M. Volatile components of an extract composition. J. Food Sci., 1985, 50: 125~131

[27] Werkhoff P, Brüning J, Emberger R et al. Studies on volatile sulphur-containing flavour

components in yeast extract//Bhattacharyya S C, Sen N, Sethi K L. 11th International Congress of Essential Oils, Fragrances and Flavours. Proceedings: Volume 4, Chemistry Analysis and Structure. New Delhi: Oxford & Ibh publishing Co. Pvt. Ltd., 1989. 215~243

[28] Hartman G J, Carlin J T, Scheide J D et al. Volatile products formed from the thermal degradation of thiamin at high and low moisture levels. J. Agric. Food Chem., 1984, 32 (5): 1015~1018

[29] 谢建春,孙宝国,刘玉平等. 维生素 B_1 微波加热降解香味成分分析. 食品科学,2004, 25 (10):241~244

[30] Werkhoff P, Brüning J, Emberger R et al. Isolation and characterization of volatile sulfur-containing meat flavor components in model systems. J. Agric. Food Chem., 1990, 38(3): 777~791

[31] 谭斌,丁霄霖. 模式体系 Maillard 反应肉类(牛肉)香精的挥发性成分分析. 香料香精化妆品,2005,(1):9~14

[32] 郭新颜,孙宝国,宋焕禄. 在 Maillard 反应的鸡肉味香精中二(2-甲基-3-呋喃基)-二硫醚的鉴别. 化学通报,2001,(1): 64~66

[33] 孙宝国,田红玉,郑福平等. 中国含硫香料的现状和发展趋势//中国香料香精化妆品工业协会. 2006 年中国香料香精学术研讨会论文集. 北京:中国香料香精化妆品工业协会,2006. 99~102

[34] Newberne P, Smith R L, Doull J et al. GRAS Flavoring substances 19. Food Technology, 2000, 54(6): 66~84

[35] 张克强,孙宝国. 1-(并二甲硫基)甲基-4-甲氧基苯的合成. 精细化工,1997,14(4):25~27

[36] 张克强,孙宝国. 苯甲醛正丁硫醇缩醛的合成研究. 精细化工,1998,15(3):23~25

[37] 孙宝国,杨迎庆,郑福平等. 15 种 1,3-氧硫杂环戊烷类香料化合物的合成. 化学通报, 2002,65(9):614~619

第二章 含硫化合物分子结构与香味的关系

香料化合物分子结构与香味的关系一直是香味化学研究的热点和难点。由于香料分子多样性和结构复杂性的影响,更由于人描述香味的生理过程的复杂性的影响,在分子结构和香味之间确定一种能肯定地预测某种新化合物香味特征的理论,到目前为止还没有成功。迄今为止,已经取得的成果仅局限于某类特定结构的化合物分子结构与香味的关系上,这种关系大都建立在对经验总结的基础之上,虽然还不能从理论的高度加以解释,但对于香料新化合物的合成和调香却有很大的指导作用。例如,C_{14}~C_{18} 的大环酮具有麝香香气;具有如图 2-1 所示结构的环酮类化合物具有焦糖香味[1]。

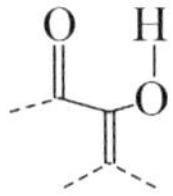

图 2-1 焦糖香味化合物的特征分子骨架

在本章中,我们重点对肉香味、葱蒜香味、咖啡香味含硫化合物分子结构与香味的关系进行论述。

第一节 肉香味含硫化合物分子结构与香味的关系

一、3-呋喃硫化物和 3-噻吩硫化物

含硫化合物是肉香味的核心和基础,肉香味含硫化合物分子结构与香味关系的研究一直是研究的热点。Dimoglo 等经过研究发现了如下规律:具有如图 2-2 所示结构单元的香料一般具有肉香味[2]。

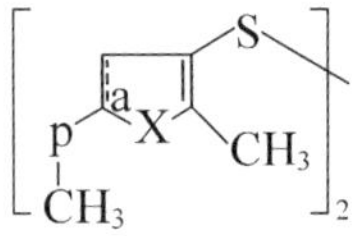

图 2-2 肉香味香料结构单元

X 为 O 或 S;p 为与 a 碳原子和甲基碳原子共面的基团(如 C=O),或不同于 O 的原子(如 S)

对于呋喃和噻吩的衍生物,2-位甲基对化合物的肉香味有重要作用,甲基必须能自由旋转。另外,当 3-位碳上的硫原子所连基团为呋喃环或含两个碳原子以上

的基团更有利于肉香。同时,Dimoglo等还指出甲基上的氢原子与X原子之间的距离也很重要。上述发现对于3-呋喃硫化物系列香料的研究开发和应用起到了非常大的促进作用,该类香料已经发展成为最重要的一类肉香味香料,并且还继续有新的品种被批准使用[3,4]。

Dimodo等的发现局限于3-呋喃硫化物和3-噻吩硫化物。

2001年,我们在对具有FEMA号的29种含硫香料香味特征归纳总结的基础上,首次提出了肉香味含硫化合物的特征分子骨架,指出含有如图2-3所示结构的有机化合物都具有基本肉香味[5]。所涉及的香料有2-甲基-3-呋喃硫醇系列、α,β-二硫系列、3-巯基-2-丁醇、α-巯基酮系列、1,4-二噻烷系列、四氢噻吩-3-酮系列和糠硫醇系列七类。在此后的研究中,我们以巯基乙醇和羰基化合物为原料,以对甲苯磺酸为催化剂,在苯中共沸脱水合成了一系列符合如图2-3所示结构的1,3-氧硫杂环戊烷类化合物[6~9],并对这些化合物的香气特征进行了评价,发现这些化合物都具有肉类香气特征[10]。

S〜X

图2-3 肉香味含硫化合物的特征分子骨架

X为O或S

2003年,我们进一步将如图2-3所示肉香味含硫化合物特征分子骨架分解为如图2-4所示的6种形式,对已有香气、香味报道的含有这些特征结构单元的化合物的香味特征进行了分类、归纳和总结。研究结果表明,它们都具有基本肉香味特征[11]。

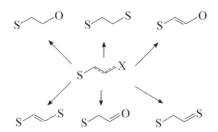

图2-4 肉香味含硫化合物分子的6种特征结构单元

二、含有各种特征结构单元的肉香味含硫化合物

1. 含有一个S—C—C—O特征结构单元的化合物

3-巯基-2-丁醇是此类化合物的典型代表,具有肉香和烤肉香[5],是肉味香精中常用的香料,2-甲基-3-四氢呋喃硫醇也具有肉香和烤肉香[5],糠硫醇、硫代乙酸糠

酯、硫代丙酸糠酯都具有肉香[5]。其他一些含有一个 S—C—C—O 特征结构单元的化合物的香味特征见表 2-1。

表 2-1 含有一个 S—C—C—O 特征结构单元的化合物的香味特征

名称	化学结构式	香味特征
2-乙基-5-甲基-3-四氢呋喃硫醇		肉香、烤肉香[12]
2-甲基-3-(2,5-二甲基-2-四氢呋喃硫基)四氢呋喃		牛肉香[12]
5-甲基糠硫醇		肉香[13]
4-甲基-4-糠硫基-2-戊酮		烤肉香[14]
糠基 2-吡嗪甲基硫醚		煮肉香、咖啡香、巧克力香[15]
2-甲基-3-糠硫基吡嗪		煮肉香、咖啡香[15]
2-甲基-5-糠硫基吡嗪		煮肉香、咖啡香[15]
硫代糠酸甲酯		肉香、海鲜香[16]
1,3-氧硫杂环戊烷		洋葱香、大蒜香、烤肉香[10]
2-甲基-1,3-氧硫杂环戊烷		洋葱香、大蒜香、烤肉香[10]
2-乙基-1,3-氧硫杂环戊烷		洋葱香、大蒜香、烤肉香[10]
2,5-二甲基-1,3-氧硫杂环戊烷		洋葱香、大蒜香、烤肉香[10]

续表

名称	化学结构式	香味特征
2-甲基-2-乙基-1,3-氧硫杂环戊烷		洋葱香、大蒜香、烤肉香[10]
2-丙基-1,3-氧硫杂环戊烷		烤肉香、洋葱香、大蒜香、辛香[10]
2-丙烯基-1,3-氧硫杂环戊烷		烤肉香、青叶香[10]
2-甲基-2-丙基-1,3-氧硫杂环戊烷		洋葱香、烤肉香[10]
2-丁基-1,3-氧硫杂环戊烷		洋葱香、大蒜辛香、烤肉香[10]
2-甲基-2-戊基-1,3-氧硫杂环戊烷		芹菜香、烤肉香、洋葱香、大蒜香[10]
2-己基-1,3-氧硫杂环戊烷		草莓香、烤肉香、硫磺样[10]
2-庚基-1,3-氧硫杂环戊烷		芹菜香、烤鸡香[10]
2-巯基-1-乙醇缩环戊酮		洋葱香、大蒜香、肉香[10]
2-巯基-1-乙醇缩乙酰乙酸乙酯		苹果香、草莓香、火腿香[10]
2-甲基-2-丙酮烷基-1,3-氧硫杂环戊烷		药味、烤肉香[10]
2-苯基-1,3-氧硫杂环戊烷		烤肉香、杏仁香[10]
2-甲基-2-苯基-1,3-氧硫杂环戊烷		杏仁香、胡桃香、松仁香、烤肉香[10]

续表

名称	化学结构式	香味特征
2-乙基-2-苯基-1,3-氧硫杂环戊烷	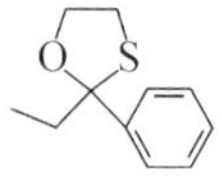	杏仁香、烤肉香、硫磺样[10]
4-[2-(1,3-氧硫杂环戊基)]丁醛	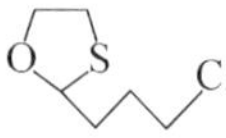	烤肉香[10]

表 2-1 中含有一个 S—C—C—O 特征结构单元的含硫化合物都具有基本肉香味特征,符合图 2-3 的结论。

2. 含有一个 S—C—C—S 特征结构单元的化合物

这类含硫化合物中具有基本肉香味的除了 1,2-乙二硫醇、2,3-丁二硫醇、α-甲基-β-羟基丙基 α'-甲基-β'-巯基丙基硫醚、1,4-二噻烷等[5]外,其他一些含有一个 S—C—C—S 特征结构单元的化合物的香味特征见表 2-2。

表 2-2 含有一个 S—C—C—S 特征结构单元的化合物的香味特征

名称	化学结构式	香味特征
3-丙硫基-2-丁硫醇	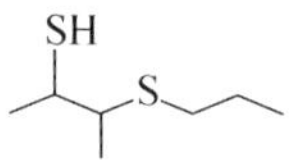	烤肉香[12]
2-甲基-3-四氢噻吩硫醇	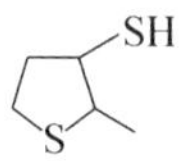	肉香[12,17,18]
2-甲基-4-四氢噻吩硫醇	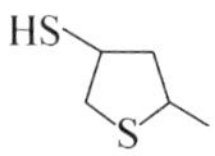	肉香[17,18]
2-甲基-3-(2-甲基-2-四氢呋喃硫基)四氢噻吩	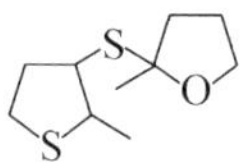	牛肉香[12]
2-甲基-2,3-二氢-3-二氢噻吩硫醇	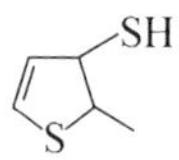	甜的、肉香、烤肉香[12,15,17]
2-甲基-4,5-二氢-4-二氢噻吩硫醇	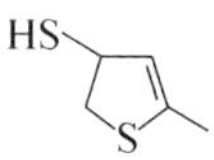	烤肉香[12,15,19]
2-甲基-1,3-二硫杂环戊烷	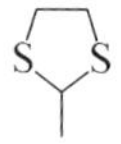	烤肉香[18,20]

续表

名称	化学结构式	香味特征
2,3-丁二硫醇缩 3-甲硫基丙醛		肉香、HVP[1)]样、肉提取物样香气和味道[21]
3-甲基-1,2,4-三硫杂环己烷		烤香、烤肉香[12,20,22]

1) HVP 为水解植物蛋白(hydrolyzed vegetable protein)的英文缩写。

表 2-2 中含有一个 S—C—C—S 特征结构单元的含硫化合物都具有基本肉香味特征,符合图 2-3 的结论。

3. 含有一个 S—C=C—O 特征结构单元的化合物

所有的 3-呋喃硫化物都含有 S—C=C—O 结构单元,都具有肉香味。前面已论述过 12 个具有肉香味的该类化合物,其中 2-甲基-3-呋喃硫醇、2-甲基-3-甲硫基呋喃、甲基 2-甲基-3-呋喃基二硫醚、二(2-甲基-3-呋喃基)二硫醚是最重要的肉香味香料[5]。其他一些含有一个 S—C=C—O 特征结构单元的化合物的香味特征见表 2-3。

表 2-3 含有一个 S—C=C—O 特征结构单元的化合物的香味特征

名称	化学结构式	香味特征
2-甲基-4,5-二氢-3-呋喃硫醇		肉香、烤肉香[12,17,20]
2-甲基-4-呋喃硫醇		青香、肉香、烤肉香[12,17]
3-乙酰硫基呋喃		肉香[12]
2-甲基-3-乙酰硫基呋喃		肉香[12,23]
2,5-二甲基-3-乙酰硫基呋喃		烤肉香[24]
2,5-二甲基-3-(2-甲基丙酰硫基)呋喃		肉香、坚果香[24]

续表

名称	化学结构式	香味特征
2-甲基-3-丁酰硫基呋喃		甜的、烤肉香[25]
2-甲基-3-(3-甲基丁酰硫基)呋喃		肉香[24]
2,5-二甲基-3-(2-甲基丁酰硫基)呋喃		肉香、烤肉香气和味道[24]
2-甲基-3-(2-乙基丁酰硫基)呋喃		肉香[24]
2,5-二甲基-3-(2-乙基丁酰硫基)呋喃		甜的、坚果香、肉香[12,24]
2-甲基-3-(2,2-二甲基丙酰硫基)呋喃		甜的、烤肉香[24]
2,5-二甲基-3-(2,2-二甲基丙酰硫基)呋喃		甜的、肉香[24]
2,5-二甲基-3-(2-甲基-2-戊烯酰硫基)呋喃		肉香[24]
2-甲基-3-糠酰硫基呋喃		HVP样、肉香[12,24]
2,5-二甲基-3-糠酰硫基呋喃		肉香[24]
2,5-二甲基-3-己酰硫基呋喃		肉香[24]
2-甲基-3-辛酰硫基呋喃		肉香[24]
2,5-二甲基-3-辛酰硫基呋喃		肉香、肉汤香、坚果香[24]

续表

名称	化学结构式	香味特征
2,5-二甲基-3-苯甲酰硫基呋喃		煮鸡香、烤肉香[24]
2,5-二甲基-3-肉桂酰硫基呋喃		甜的、肉香[24]
2,5-二甲基-3-(3-甲基苯甲酰硫基)呋喃		肉香、烤肉香[24]
2-甲基-3-(2-硫杂环丁烷硫基)呋喃		硫磺样、甘蓝香、肉香、橡胶样[25]
2-甲基-3-(2-甲基-2-四氢噻吩硫基)呋喃		硫磺样、烤香、肉香[26]
3-(2-甲基-2,3-二氢呋喃-2-硫基)呋喃		牛肉香[14]
3-呋喃基 2-莰基硫醚		肉香[14]
咖啡呋喃		煮肉香、烤肉香、烟熏香[13,19]
3-(2-甲基烯丙硫基)呋喃		煮肉香、牛肉汤香、肉香、坚果香[14,27]
2,5-二甲基-3-丙硫基呋喃		肉香、烤肉香、烤牛肉香[12,24]
2-甲基-3-(3-甲基丁硫基)呋喃		甜的、烤面包香、肉香、烤牛肉香[12,28]
2-甲基-3-(2-丁烯-1-硫基)呋喃		橡胶样、硫磺样、肉香[27]
2-甲基-3-环己硫基呋喃		肉香、烤牛肉香[12]

续表

名称	化学结构式	香味特征
2-甲基-3-苄硫基呋喃	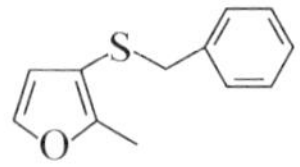	甜的、烤香、肉香[27]
2-甲基-3-(2-吡啶甲硫基)呋喃	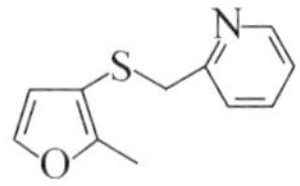	肉香、牛肉香[14]
2-甲基-3-(1-巯基乙硫基)呋喃	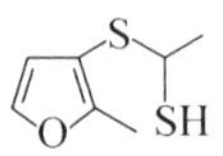	烤香、肉汤香、辛香、洋葱香、大蒜香、蔬菜香、肉香、肉汁香、烤牛肉香[24]
甲基 2,5-二甲基-3-呋喃基二硫醚	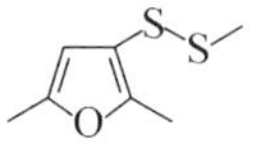	甜的、烤肉香[29]
乙基 2-甲基-3-呋喃基二硫醚	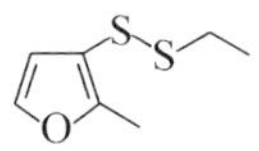	甜的、烤香、肝香、烤牛肉香[29]
2-甲基-3-(2-硫杂丙硫基)呋喃	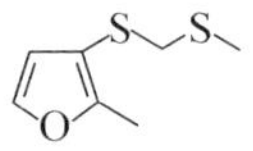	麦香、青香、硫磺样、甜的、肉香[27]
丙基 2,5-二甲基-3-呋喃基二硫醚	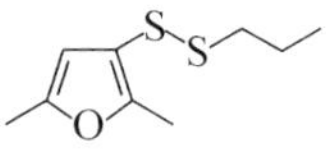	甜的、烤香、肉香、HVP样[29]
异戊基 2-甲基-3-呋喃基二硫醚	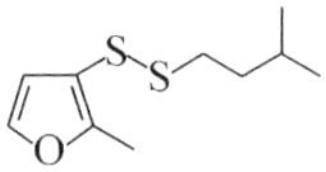	肉香、肝香、鸡脂香、甜的、坚果香、烤香[29]
环己基 2-甲基-3-呋喃基二硫醚	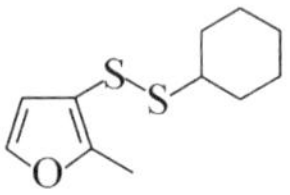	肉香、肝香、甜的、坚果香[29]
甲基 2-甲基-3-呋喃基三硫醚	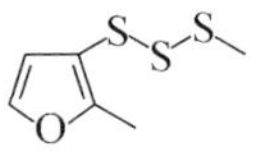	肉香[25]

表 2-3 中含有一个 O—C═C—S 特征结构单元的含硫化合物都具有基本肉香味特征，符合图 2-3 的结论。

4. 含有一个 S—C═C—S 特征结构单元的化合物

所有的 3-噻吩硫化物都含有 S—C═C—S 结构单元，都具有肉香味。前面已论述过该类化合物中的 2-甲基-3-甲硫基噻吩、甲基 2-甲基-3-噻吩基二硫醚等[5]，但这类化合物尚未获得有关组织的批准，还不允许在香精中使用。其他一些含有一个S—C=C—S特征结构单元的化合物的香味特征见表2-4。

表 2-4 含有一个 S—C=C—S 特征结构单元的化合物的香味特征

名称	化学结构式	香味特征
3-噻吩硫醇		脂肪香、洋葱香、咖啡香、煮肉香[18]
2-甲基-3-噻吩硫醇		肉香、烤肉香[12,17]
2-甲基-2,3-二氢-4-噻吩硫醇		肉香[12,17,18]
2-甲基-4,5-二氢-4-噻吩硫醇		烤肉香[18]
2-甲基-4,5-二氢-3-噻吩硫醇		肉香[12,17,18]
2-甲基-2,3-二氢-3-噻吩硫醇		甜的、烤肉香[18]
2-甲基-3-(2-甲基-2-四氢噻吩硫基)噻吩		洋葱香、甘蓝香、烤肉香[26]
甲基 2-甲基-3-噻吩基三硫醚		肉香、烤香[25]

表 2-4 中含有一个 S—C=C—S 特征结构单元的含硫化合物都具有基本肉香味特征,符合图 2-3 的结论。

5. 含有一个 S—C—C=O 特征结构单元的化合物

这类化合物中具有基本肉香味的 1-巯基-2-丙酮、3-巯基-2-丁酮、3-巯基-2-戊酮、3-四氢噻吩酮和 2-甲基-3-四氢噻吩酮前面已有论述[5],其他一些含有一个 S—C—C=O 特征结构单元的化合物的香味特征见表 2-5。

表 2-5　含有一个 S—C—C═O 特征结构单元的化合物的香味特征

名称	化学结构式	香味特征
巯基丙酮		肉香[12]
3-甲基-3-巯基戊酮		肉香[12]
2-巯基丁二醛		烤肉香[12]
3-巯基-2,5-己二酮		烤肉香[12]
3-乙酰硫基-2,5-己二酮		烤肉香[12]
3-苯甲酰硫基-2,5-己二酮		烤肉香[12]
1-(2-巯基乙酰基)-3(或4)-羟基苯		肉香[12]
1-(2-巯基乙酰基)-3(或4)-甲基苯		肉香[12]
1-(2-巯基乙酰基)-3(或4)-甲氧基苯		肉香[12]
1-(2-巯基乙酰基)-3(或4)-苄基苯		肉香[12]
1-丙硫基-1,3-二苯基丙酮		肉香、洋葱香[14]

续表

名称	化学结构式	香味特征
2-噻吩甲醛	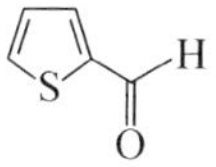	辛香、肉香、坚果香[18,20]
3-乙酰基-1,3-二硫杂环戊烷	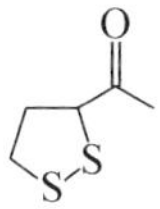	肉香、洋葱香、肝香[18]
2-甲基-3,4-二硫杂环己酮	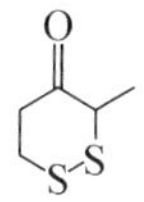	血腥味、硫磺样、肉香[13]
2-乙酰基噻唑	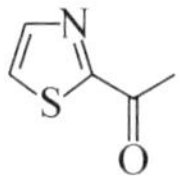	牛肉香[22]
5-乙酰基噻唑	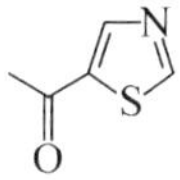	烤肉香[22]
2,4-二甲基-5-乙酰基噻唑	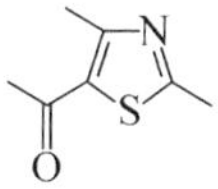	霉香、坚果香、壤香、咖啡香、肉香[22,30]
2-乙酰基-2-噻唑啉	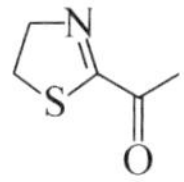	烤牛肉香[22]

表 2-5 中含有一个 S—C—C＝O 特征结构单元的含硫化合物都具有基本肉香味特征,符合图 2-3 的结论。

6. 含有多个特征结构单元的化合物

已经允许使用的含有多个特征结构单元的含硫香料如二糠基二硫醚、二(2-甲基-3-呋喃基)二硫醚、二(2,5-二甲基-3-呋喃基)二硫醚、二(2-甲基-3-呋喃基)四硫醚、2,5-二甲基-3-糠酰硫基呋喃、2,5-二羟基-1,4-二噻烷、2,5-二甲基-2,5-二羟基-1,4-二噻烷、α-甲基-β-羟基丙基 α′-甲基-β′-巯基丙基硫醚、螺[2,4-二硫杂-1-甲基-8-氧杂二环[3.3.0]辛烷-3,3′-(1′-氧杂-2′-甲基)环戊烷]和螺[2,4-二硫杂-6-甲基-7-氧杂二环[3.3.0]辛烷-3,3′-(1′-氧杂-2′-甲基)环戊烷]等都具有基本肉香味[5],其他一些含有多个特征结构单元的化合物的香味特征见表 2-6。

表 2-6 含有多个肉香味特征结构单元的化合物及其香味特征

名称	化学结构式	香味特征
2-甲基-2-乙酰基-1,3-氧硫杂环戊烷		鸡肉香、煮蘑菇香[10]
2-(2-呋喃基)-1,3-氧硫杂环戊烷		煮肉香、咖啡香[10]
1,3-二(1,3-氧硫杂环戊-2-基)丙烷		烤肉香、干蘑菇香[10]
4-甲基-2-(2-硫杂丙基)-1,3-二硫杂环戊烷		肉香、洋葱香、烤香、HVP 样[21]
4,5-二甲基-2-(2-硫杂丙基)-1,3-二硫杂环戊烷		甜的、肉香、HVP 样[21]
2,5-二甲基-3-(3-巯基-2-丁硫基)呋喃		牛肉汤香、火腿香、肉香、坚果香[12,14]
2,5-二甲基-3-(3-羟基-2-丁硫基)呋喃		肉香、坚果香[12]
2,5-二甲基-3-(2-羟基环己硫基)呋喃		肉香、坚果香[12]
丙酮烷基 2-甲基-3-呋喃基二硫醚		烤肉香、猪肉香、洋葱香[25]
3-戊酮-2-烷基 2-甲基-3-呋喃基二硫醚		肉香[25]
二糠基硫醚		牛肉香、鸡肉香[19]
二(3-噻吩基)二硫醚		烤肉香[12]

续表

名称	化学结构式	香味特征
二(2-甲基-3-噻吩基)二硫醚		硫磺样、金属样、橡胶样、淡肉香[18]
二(2-甲基-3-呋喃基)硫醚		烤肉香[12]
二(2,5-二甲基-3-呋喃基)硫醚		烤肉香[12]
2-甲基-3-呋喃基 2-丙基-3-呋喃基硫醚		烤肉香[12]
二(2-丙基-3-呋喃基)硫醚		烤肉香[12]
2-甲基-3-呋喃基 2-甲基-3-噻吩基二硫醚		洋葱香、大蒜香、肉香、金属样、脂肪香[13,18,20]
2-甲基-3-呋喃基 2-甲基-3-四氢噻吩基硫醚		坚果香、蘑菇香、面包香、肉香、烤肝香[26]
2-甲基-3-噻吩基 2-甲基-3-四氢噻吩基硫醚		蘑菇香、肉香、烤肉香[26]
2-甲基-3-(2-四氢噻吩甲硫基)呋喃		硫磺样、韭葱香、细香葱香、大蒜香、洋葱香、淡肉香[26]
2-甲基-3-(2-四氢噻吩甲硫基)噻吩		硫磺样、橡胶样、洋葱香、淡肉香[26]
		硫磺样、肉香、花生香、豆香[25]
		土豆香、硫磺样、肉香、甘蓝香[25]

续表

名称	化学结构式	香味特征
2-甲基-3-巯基-3-(2-甲基-4,5-二氢-3-呋喃硫基)四氢呋喃		肉香[12]
2-甲基-3-巯基-3-(2-甲基-2,5-二氢-3-呋喃硫基)四氢呋喃		肉香[12]
2,5-二甲基-3-巯基-3-(2,5-二甲基-2,5-二氢-3-呋喃硫基)四氢呋喃		肉香[12]
4-巯基-3-四氢呋喃酮		青香、肉香、烤肉香、肉汁样[12,17]
5-甲基-4-巯基-3-四氢呋喃酮		肉香、肉汁样[17,18]
5-甲基-4-羟基-3-四氢噻吩酮		烤肉香[22]
5-甲基-4-羟基-2,3-二氢噻吩-3-酮		烤肉香[22]
5-甲基-4-巯基-3-羟基-4,5-二氢噻吩		肉香[17]
2,5-二乙基-2,5-二羟基-1,4-二噻烷		肉香[12]
		肉香、牛肉香[12]
		肉香、牛肉香[12]

续表

名称	化学结构式	香味特征
	(结构式图)	肉香、牛肉香[12]
	(结构式图)	肉香[12]
二(2-甲基-3-巯基丙基)硫醚	(结构式图)	烤肉香[14]
3-硫杂-4-乙基-5-氧代辛酸甲酯	(结构式图)	甜的、肉香[14]
D-葡萄糖二甲硫醇缩醛	(结构式图)	肉香[12,31]

表 2-6 中含有多个特征结构单元的含硫化合物都具有基本肉香味特征,符合图 2-3 的结论。

含有一个和多个 S—C—C—O、S—C—C—S、S—C=C—O、S—C=C—S、S—C—C=O 特征结构单元的化合物都具有基本肉香味特征,符合图 2-3 的结论。这一结论用来指导肉香味含硫新化合物的合成,可以避免分子设计的盲目性,提高肉香味化合物筛选的成功率;用来指导肉味香精调香,能提高肉香味香料选择的准确性。含有 S—C—C=S 特征结构单元化合物的香味特征,目前尚未发现有关评价的报道,是否符合图 2-3 的结论还需要进一步研究。

第二节 葱蒜香味含硫化合物分子结构与香味的关系

葱蒜香味是包括洋葱、大葱、韭葱、细香葱、大蒜、韭菜等香味的总称。具有丙硫基、烯丙硫基和丙烯硫基基团的化合物一般具有葱蒜香味,这类化合物可以用如

图 2-5 所示结构通式表示。

$$H_3C\diagdown\diagup SR$$

图 2-5 葱蒜香味含硫化合物的结构通式

如图 2-5 所示的化合物中含有烯丙硫基的化合物其大蒜香味特征更突出,而含有丙硫基的化合物其洋葱香味特征更明显[32]。

符合这一规律的葱蒜香味香料有二丙基硫醚、甲基丙基二硫醚、二丙基二硫醚、甲基丙基二硫醚、二丙基三硫醚、甲基丙基三硫醚、烯丙硫醇、烯丙基硫醚、甲基烯丙基硫醚、丙基烯丙基硫醚、烯丙基二硫醚、甲基烯丙基二硫醚、丙基烯丙基二硫醚、烯丙基三硫醚、甲基烯丙基三硫醚、丙基烯丙基三硫醚以及二烯丙基二硫醚和丙基烯丙基二硫醚等。部分葱蒜香味化合物的化学结构和香味特征列于表 2-7 中。

表 2-7 部分葱蒜香味化合物的化学结构及其香味特征

名称	化学结构式	香味特征
丙硫醇	 ⌒⌒SH	洋葱及甘蓝香气[33]
二丙基硫醚	 ⌒⌒S⌒⌒	葱香、蒜香香气[33]
1-丙硫基-2-丁醇	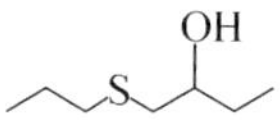 ⌒⌒S-CH(OH)-⌒	葱蒜香气、带青香
1-烯丙硫基-2-丁醇	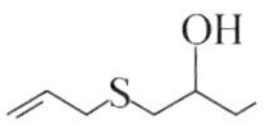 CH2=CH-CH2-S-CH(OH)-⌒	腌制过的萝卜香气、带葱蒜香气
1-烯丙硫基-2-丁酮	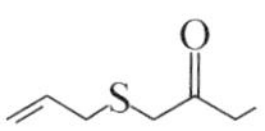 CH2=CH-CH2-S-C(=O)-⌒	0.01%时具有微弱的葱蒜香气、似萝卜香气
二丙基二硫醚	 ⌒⌒SS⌒⌒	洋葱香、大蒜香气[33]
二丙基三硫醚	 ⌒⌒SSS⌒⌒	洋葱香、大蒜样香[33]
甲基丙基二硫醚	 CH_3SS⌒⌒	硫磺样、葱香、蒜香、小萝卜香、芥菜香、西红柿香、土豆香、大蒜香气[33]
甲基丙基三硫醚	 CH_3SSS⌒⌒	洋葱香味[32]

续表

名称	化学结构式	香味特征
烯丙硫醇	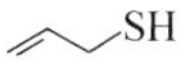 ⌒⌒SH	强烈的大蒜气味[33]
二烯丙基硫醚	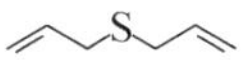 ⌒S⌒	洋葱香、大蒜香、蔬菜香、萝卜香、辣根样香韵[33]
二烯丙基二硫醚	 ⌒SS⌒	大蒜香味[32]
二烯丙基三硫醚	 ⌒SSS⌒	大蒜香味[32]
丙基烯丙基二硫醚	 ⌒SS⌒	强烈的蒜头香气[33]
甲基烯丙基二硫醚	 CH_3SS⌒	洋葱香、韭菜香、大蒜香及盐渍的蒜香香气[33]
甲基烯丙基三硫醚	 CH_3SSS⌒	大蒜香、洋葱香气和味道[33]
丙基丙烯基二硫醚	 ⌒SS⌒	烹煮洋葱似的香气和味道[33]
二丙烯基二硫醚	 ⌒SS⌒	大蒜香味[32]
硫代丙酸烯丙酯	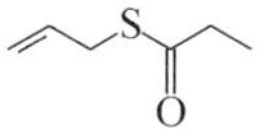 ⌒S-C(=O)-⌒	青香及洋葱、大蒜的气息[33]
硫代乙酸丙酯	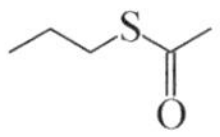 ⌒S-C(=O)-CH₃	硫磺样、生洋葱香、大蒜香、蔬菜香气[33]

第三节 咖啡香味含硫化合物分子结构与香味的关系

含有糠硫基的化合物一般具有咖啡或芝麻油(香油)香味,这类化合物可以用如图 2-6 所示的结构通式表示。

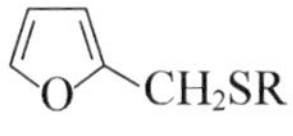

$$\text{（呋喃环）}—CH_2SR$$

图 2-6 咖啡香味含硫化合物的结构通式

符合上述结论的化合物及其香味特征列于表 2-8 中。

表 2-8 咖啡香味化合物及其香味特征

名称	化学结构式	香味特征
糠硫醇	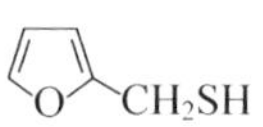 呋喃-CH$_2$SH	芝麻香、洋葱香、大蒜香、肉香及咖啡香气[33]
糠基异丙基硫醚	 呋喃-CH$_2$SCH(CH$_3$)$_2$	葱蒜香、咖啡香气[33]
3-糠硫基丙酸乙酯	 呋喃-S-CH$_2$CH$_2$COOC$_2$H$_5$	洋葱香、大蒜香、咖啡香、蔬菜香、烤香、坚果香[33]
甲基糠基二硫醚	 呋喃-CH$_2$SSCH$_3$	蔬菜香、洋葱香、辣根香、咖啡香、硫磺样香气[33]
二糠基二硫醚	 呋喃-CH$_2$SSCH$_2$-呋喃	咖啡香、烤香、肉香、洋葱香、甘蓝香、土豆香[33]
硫代乙酸糠酯	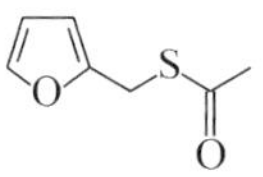 呋喃-CH$_2$-S-C(=O)CH$_3$	蔬菜香、硫磺、烤香、大蒜香、咖啡香、肉香[33]
硫代丙酸糠酯	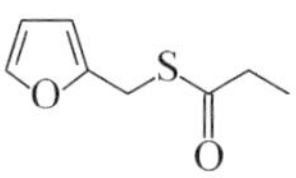 呋喃-CH$_2$-S-C(=O)CH$_2$CH$_3$	硫磺样、咖啡香、洋葱香、大蒜香、肉香、烤香[33]

含硫香料分子结构与香味之间的关系是非常复杂的,既有分子结构本身的原因,也有浓度、温度等因素的影响。同一种含硫香料在不同情况下(如浓度)可能表现出不同的香味特征。同一种香料分子中含有不同的特征结构单元可能表现出不同的香味特征,如糠硫醇、二糠基二硫醚、硫代乙酸糠酯等分子结构中同时含有如图 2-3、图 2-5、图 2-6 所示的结构单元,具有肉香、葱蒜香、咖啡香等香味特征。

参 考 文 献

[1] 〔日〕精细化学品辞典编辑委员会. 精细化学品辞典. 禹茂章等译. 北京:化学工业出版社,1989. 807~848

[2] Dimoglo A S, Gorbachov M Y, Bersuker I B et al. Structural and electronic origin of meat odour of organic heteroatomic compounds. Nahrung, 1988, 32: 461~473

[3] Newberne P, Smith R L, Doull J et al. GRAS flavoring substances 19. Food Technology, 2000, 54(6): 66~84

[4] Smith R L, Doull J, Feron V J et al. GRAS flavoring substances 20. Food Technology,

2001,55(12):34～55
- [5] 孙宝国,丁富新,郑福平等. 含硫香料分子结构与肉香味的关系研究. 精细化工,2001,18(8):456～460
- [6] 杨迎庆,郑福平,张军,孙宝国等. 2-甲基-1,3-氧硫杂环戊烷和 2-丙基-1,3-氧硫杂环戊烷的合成. 精细化工,2001,18(9):525～527
- [7] 孙宝国,杨迎庆,郑福平等. 15种1,3-氧硫杂环戊烷类香料化合物的合成. 化学通报,2002,65(9):614～619
- [8] Yang Y Q, Zheng F P, Sun B G et al. Synthesis of three 2-alkyl(aryl)-1,3-oxathiolane compounds. Chemical Journal on Internet,2001,3(11)
- [9] Zheng F P, Sun B G, Liu Y P et al. Syntheses of six novel 1,3-oxathiolane flavoring compounds//Yang J Z, Peng X J. The proceedings of The 2nd international conference on functional molecules. Dalian:Dalian University of Technology Press, 2003. 284～286
- [10] Sun B G, Miao C F, Yang Y Q et al. Aroma characteristics of 1,3-oxathiolanes// Yang J Z, Peng X J. The proceedings of The 2nd international conference on functional molecules. Dalian:Dalian University of Technology Press, 2003. 377～379
- [11] Sun B G, Tian H Y, Zheng F P et al. Characteristic structural unit of sulfur-containing compounds with a basic meat flavor. Perfumer & Flavorist, 2005, 30(1):36～45
- [12] MacLeod G, Seyyedain-Ardebili M. Natural and simulated meat flavors (with particular reference to beef). CRC Critical Reviews in Food Science and Nutrition, 1981, 14(4): 308～437
- [13] Werkhoff P, Brüning J, Emberger R et al. Studies on volatile sulphur-containing flavour components in yeast extract//Bhattacharyya S C, Sen N, Sethi K L. 11th International Congress of Essential Oils, Fragrances and Flavours. Proceedings:Volume4, Chemistry Analysis and Structure. New Delhi:Oxford & Ibh publishing Co. Pvt. Ltd., 1989. 215～243
- [14] 黄荣初,王兴凤. 肉类香味的合成香料. 有机化学,1983,(3):175～179
- [15] Calabretta P. Synthesis of some substituted pyrazines and their olfactive properties. Perfumer & Flavorist, 1978, (3):33～42
- [16] 孙宝国,郑福平,刘玉平. 香料与香精. 北京:中国石化出版社,2000.152～153
- [17] Godefridus A M, Ouweland V D, Henricus G P. Components contributing to beef flavor. volatile compounds produced by the reaction of 4-hydroxy-5-methyl-3(2H)-furanone and its thio analog with hydrogen sulfide. J. Agric. Food Chem.,1975,23(3):501～505
- [18] Macleod G. The flavor of beef//Shahidi F. Flavor of Meat, Meat Products and Seafoods. London:Blackie Academic & Professional, 1998. 28～60
- [19] 陈煜强,刘幼君. 香料产品开发与应用. 上海:上海科学技术出版社,1994. 381～400
- [20] Macleod G. The flavor of beef//Shahidi F. Flavor of Meat and Meat products. New York:Blackie Academic & Professional, 1994. 4～33
- [21] Pittet A O, Courtney T F, Muralidhara R. Methyl(methylthioalkyl)-1,3-dithiolanes:

US, 4515967. 1985.5.7
- [22] Shaikh Y. Aroma chemicals in meat flavors. Perfumer & Flavorist, 1984, 9(3):49~52
- [23] Rowe D J. Aroma chemicals for savory flavors. Perfumer & Flavorist, 1998, 23(4):9~16
- [24] Evers W J, Heinsohn H R, Mayers B J. Processes for producing 3-thia furans and 3-furan thiols: US, 3922288. 1975.11.25
- [25] Mussinan C J, Keelan M E. Sulfur Compounds in Foods. Washington, DC: American Chemical Society, 1994. 185~223
- [26] Werkhoff P, Brüning J, Emberger R et al. Isolation and characterization of volatile sulfur-containing meat flavor components in model systems. J. Agric. Food Chem., 1990, 38(3):777~791
- [27] Evers W J, Heinsohn H R, Vock M H et al. Flavoring with (allyl)(2-methyl-3-furyl) sulfide: US, 4007287. 1977.2.8
- [28] Evers W J, Heinsohn H R, Mayers B J. Foodstuff flavor compositions comprising 3-furyl alkyl sulfides and processes: US, 4031256. 1977.6.21
- [29] Evers W J et al. 3-Furyl alkyl disulfides and foodstuff flavor composition comprising same: GB, 1538073. 1979.1.17
- [30] Mosciano G. Organoleptic characteristics of flavor materials. Perfumer & Flavorist, 1996, 21(6):49~52
- [31] Oftedahl M L. Flavoring food with a sugar dialkyl dithioacetals: US, 3635736. 1972.1.18
- [32] Rowe D. More fizz for your buck: High-impact aroma chemicals. Perfumer & Flavorist, 2000, 25(5):1~19
- [33] 孙宝国,刘玉平. 食用香料手册. 北京:中国石化出版社,2004. 388~419

第三章 硫醇类香料

硫醇类香料是含硫香料中最重要的一类,也是合成其他含硫香料的重要原料或中间体。其结构特点是巯基(—SH)与碳原子直接相连,结构通式为 R—SH,式中 R 为烃基或其衍生物。其中—SH 与苯环直接相连的化合物称为硫酚,在本章一并论述。

低分子的硫醇在浓度高时具有令人感觉极不愉快的气味。低浓度的硫醇类化合物常呈现令人愉快的食品类香味,如咖啡、香油、葱、蒜、萝卜、热带水果、烤肉等香味。对硫醇类化合物的香气描述中也经常出现硫磺样、橡胶样、焦糊味等字样。

硫醇类香料的品种很多,包括从结构最简单的甲硫醇到结构复杂的萜硫醇及多硫醇,目前允许作为食品香料使用的品种有 100 多个。

硫醇的沸点比相应的醇要低得多,甲硫醇常温下为气体,沸点为 5.9℃,可能是沸点最低的香料。大部分硫醇类香料常温下是液体。硫醇的酸性比醇高,如乙醇的 $pK_a=18$,而乙硫醇的 $pK_a=10.6$。硫原子的半径大于氧原子,因而易于极化,故—SH 中的氢原子易于形成氢离子,而呈现酸性。低分子的硫醇能溶于氢氧化钠的水溶液。硫醇不仅能与碱金属成盐,也可以与 Hg、Cu、Ag、Pb 等重金属形成不溶于水的硫醇盐。因此,硫醇可以作为重金属中毒的解毒剂。

$$CH_3CH_2SH + HgO \longrightarrow (CH_3CH_2S)_2Hg\downarrow + H_2O$$

硫醇很容易与氧化剂发生氧化反应,根据这一特性,可以用此反应测定巯基化合物的含量。

$$RSH + H_2O_2 \longrightarrow RSSR + H_2O$$
$$RSH + I_2 \longrightarrow RSSR + HI$$

空气中的氧在室温下也能将硫醇氧化为二硫醚。因此,硫醇类香料最好充氮气保存。

$$RSH + O_2 \longrightarrow RSSR + H_2O$$

第一节 硫醇类香料的一般合成方法

一、醇与硫化氢反应

醇与硫化氢在催化剂存在下,在气相反应生成硫醇。

$$R{-}OH + H_2S \xrightarrow{ThO_2} R{-}SH + H_2O$$

这是工业上大规模生产甲硫醇、乙硫醇、丙硫醇、丁硫醇等简单硫醇的常用方法[1]。

二、卤代物与硫化氢反应

卤代物与硫化氢在乙醚溶液中反应生成硫醇。一些 α-巯基酮类化合物可以用此方法合成。例如，α-卤代苯乙酮与硫化氢反应生成 α-巯基苯乙酮[2]。

$$C_6H_5COCH_2X + H_2S \longrightarrow C_6H_5COCH_2SH$$

又如，α-溴代戊酮与硫化氢反应生成 2-巯基-3-戊酮[3]。

三、卤代烷与硫氢化钾或硫氢化钠反应

脂肪族卤代烃很容易与硫氢化钾或硫氢化钠反应生成硫醇，这是制备硫醇类香料常用的方法之一[1]。

$$RX + KSH \xrightarrow{\triangle} RSH + KX$$

四、卤代烷和硫脲反应

卤代烷和硫脲反应先生成 s-烷基异硫脲盐，然后在碱的作用下生成硫醇[4]。

$$R{-}X + S{=}C(NH_2)_2 \longrightarrow [R{-}S{-}C(=NH)NH_2 \cdot HX] \xrightarrow{NaOH} R{-}SH$$

此方法可用于大多数硫醇化合物的合成。

五、醇和硫脲反应

醇和硫脲、氢溴酸反应先生成 s-烷基代异硫脲盐，然后在碱的作用下生成硫醇[1]。

$$R\text{—}OH + S=C(NH_2)(NH_2) \xrightarrow{HBr} [R\text{—}S\text{—}C(=NH)(NH_2 \cdot HBr)] \xrightarrow{NaOH} R\text{—}SH$$

对烯丙型醇、苯甲型醇、糠型醇、三级醇,可用浓盐酸代替氢溴酸。工业上制备糠硫醇就是采用此方法。

六、卤代烃与硫代硫酸盐反应

卤代烃与硫代硫酸盐反应先生成 Bunte 盐,后者在酸催化下水解生成硫醇[1]。

$$RX + Na_2S_2O_3 \longrightarrow NaRS_2O_3 \xrightarrow{H_2O} RSH$$

这是制备硫醇类香料常用的方法之一。

七、烯烃与硫化氢加成

烯烃与硫化氢加成生成硫醇[1]。

$$RCH=CH_2 + H_2S \longrightarrow RCH_2CH_2SH$$

这是制备某些脂肪族硫醇类香料的常用方法。

八、硫代羧酸酯水解

硫代羧酸酯在碱性条件下水解生成硫醇。例如[5]:

$$C_6H_{11}\text{—}S\text{—}C(=O)CH_3 + H_2O \xrightarrow{OH^-} C_6H_{11}\text{—}SH$$

2-甲基-3-呋喃硫醇、2,5-二甲基-3-呋喃硫醇等重要含硫香料也可采用此法合成[6]。

九、由含有环硫乙烷的化合物制备

含有环硫乙烷的化合物与硫化氢、硫氢化钠或硫氢化钾反应,这是制备 α,β-二硫醇类香料的有效方法。例如[4]:

$$\text{(环硫乙烷)} + SH^- \longrightarrow HSCH_2CH_2SH$$

十、二硫醚还原

二硫醚还原生成硫醇,这是制备某些硫醇的方法[1]。

$$RSSR + H_2 \longrightarrow RSH$$

十一、烷基氯磺酸还原

烷基氯磺酸还原生成硫醇[1]。

$$RSO_3Cl + H_2 \longrightarrow RSH$$

十二、烷基磺酸还原

烷基磺酸还原生成硫醇[1]。

$$RSO_3H + H_2 \xrightarrow{Ni} RSH$$

十三、硫酸甲酯单钠盐与硫氢化钠反应

硫酸甲酯单钠盐与硫氢化钠共热生成硫醇[1]。

$$CH_3OSO_3Na + NaSH \xrightarrow{\triangle} CH_3SH + Na_2SO_4$$

这是实验室制备甲硫醇常用的方法。

第二节　3-呋喃硫醇类香料

3-呋喃硫醇、3-二氢呋喃硫醇及 3-四氢呋喃硫醇在肉味香料中占有不可替代的重要地位,许多关键肉味香料都是它们的衍生物。

一、2-甲基-3-呋喃硫醇

2-甲基-3-呋喃硫醇(2-methyl-3-furanthiol,FEMA 号 3188)商品名"030",最早发现于金枪鱼的挥发性香成分中,以后在煮牛肉、猪肉、鸡肉及许多肉味热反应模型体系中都发现了 2-甲基-3-呋喃硫醇。该香料为浅黄色透明液体,具有肉香、鱼香、烤肉香、烤鸡香等香气以及鱼、肉、大马哈鱼、金枪鱼、烤香等味道。说 2-甲基-3-呋喃硫醇是迄今为止最重要的肉味香料是不过分的,因为目前发现的在肉香味中起关键作用的香味化合物仍是 2-甲基-3-呋喃硫醇及其衍生物。该香料可用于牛肉、猪肉、鸡肉、海鲜等咸味香精。

2-甲基-3-呋喃硫醇的合成路线主要有两条:一是以 5-甲基糠酸为原料制取;二是以 2-甲基呋喃为起始原料制取。

1. 5-甲基糠酸路线

5-甲基糠酸经磺酸化、脱羧、磺酰氯化、还原等反应,生成 2-甲基-3-呋喃硫

醇[6]。

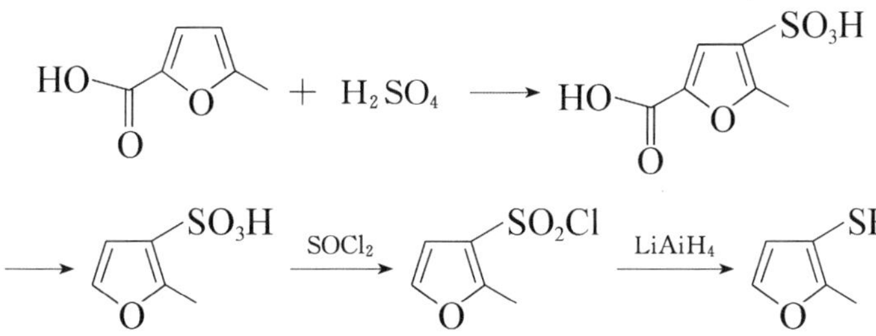

$$\text{HO-} \underset{O}{\overset{}{\text{furan-CH}_3}}\text{-COOH} + H_2SO_4 \longrightarrow \text{HO-}\underset{O}{\overset{SO_3H}{\text{furan-CH}_3}}\text{-COOH}$$

$$\longrightarrow \underset{O}{\overset{SO_3H}{\text{furan-CH}_3}} \xrightarrow{SOCl_2} \underset{O}{\overset{SO_2Cl}{\text{furan-CH}_3}} \xrightarrow{LiAlH_4} \underset{O}{\overset{SH}{\text{furan-CH}_3}}$$

2. 2-甲基呋喃路线

以 2-甲基呋喃为起始原料的合成路线需要经过以下五步化学反应：

(1) 甲氧基化反应

2-甲基呋喃与甲醇反应，生成 2-甲基-2,5-二甲基-2,5-二氢呋喃[7]。

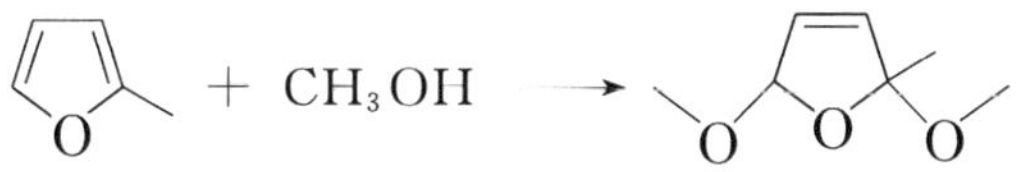

$$\underset{O}{\text{furan-CH}_3} + CH_3OH \longrightarrow \underset{\text{MeO} \ O \ \text{OMe}}{\text{dihydrofuran}}$$

2-甲基呋喃的甲氧基化反应可以在溴的作用下进行，也可以通过电解进行。

(2) 水解反应

2-甲基-2,5-二甲基-2,5-二氢呋喃在室温下水解生成 4-氧代-2-戊烯醛[8]。

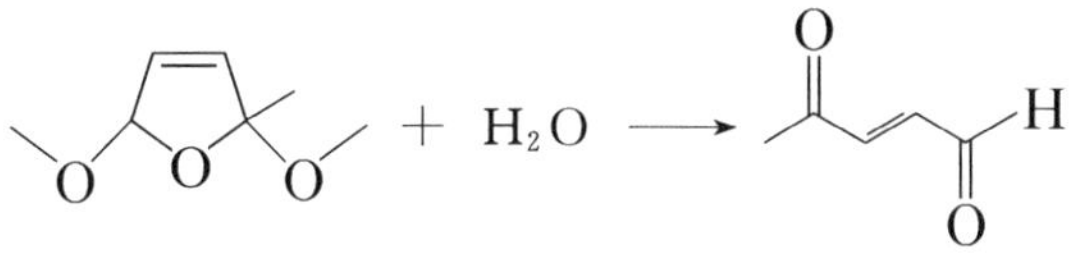

$$\underset{\text{MeO} \ O \ \text{OMe}}{\text{dihydrofuran}} + H_2O \longrightarrow \text{CH}_3\text{COCH=CHCHO}$$

(3) 加成反应

4-氧代-2-戊烯醛与硫代乙酸加成生成 3-乙酰硫基-4-氧代戊醛。

$$\text{CH}_3\text{COCH=CHCHO} + \text{CH}_3\text{COSH} \longrightarrow \text{CH}_3\text{CO-CH(SCOCH}_3\text{)-CH}_2\text{-CHO}$$

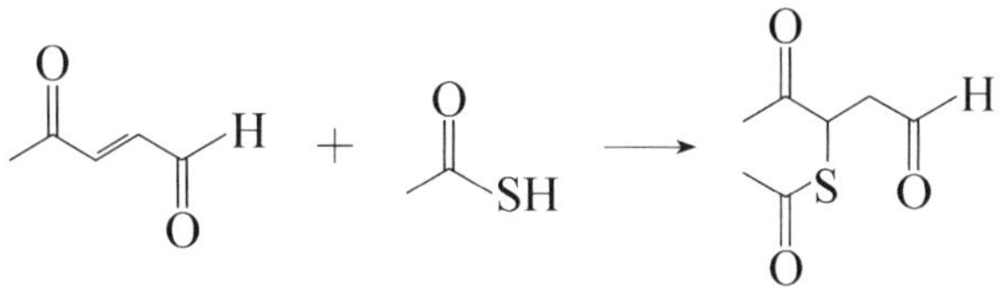

(4) 环化反应

3-乙酰硫基-4-氧代戊醛环化生成 2-甲基-3-乙酰硫基呋喃。

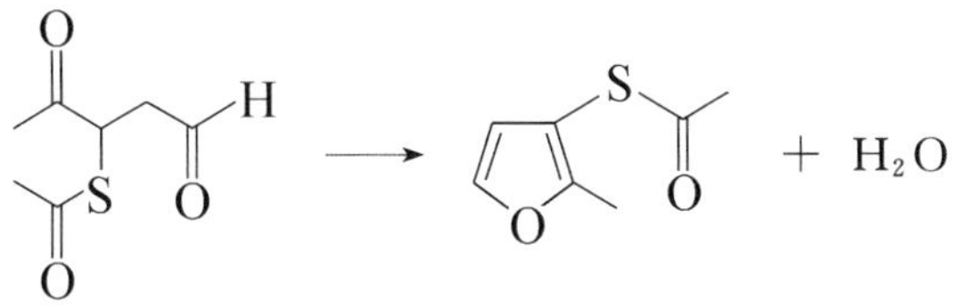

$$\text{CH}_3\text{CO-CH(SCOCH}_3\text{)-CH}_2\text{-CHO} \longrightarrow \underset{O}{\overset{SCOCH_3}{\text{furan-CH}_3}} + H_2O$$

2-甲基-3-乙酰硫基呋喃(FEMA 号 3973)也是允许使用的香料,商品名"D030"。

(5) 水解反应

2-甲基-3-乙酰硫基呋喃在碱性条件下水解生成 2-甲基-3-呋喃硫醇。

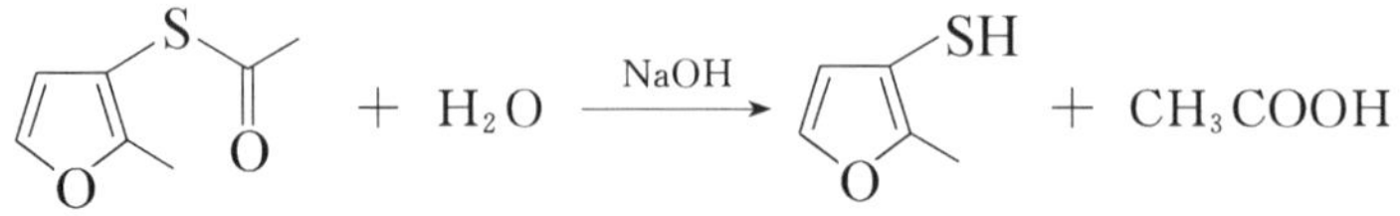

二、2,5-二甲基-3-呋喃硫醇

2,5-二甲基-3-呋喃硫醇(2,5-dimethyl-3-furanthiol,FEMA 号 3451)为浅黄色透明液体,具有强烈肉香、烤肉香,可用于各种肉味和海鲜香精。其合成方法与 2-甲基-3-呋喃硫醇类似,可采用 2,5-二甲基呋喃为原料合成[6]。

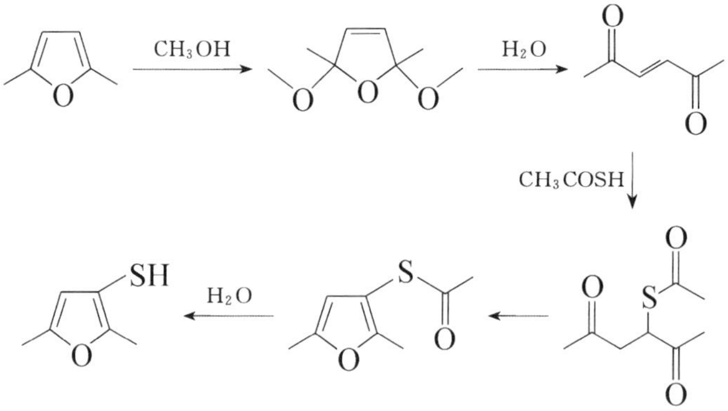

三、2-甲基-3-四氢呋喃硫醇

2-甲基-3-四氢呋喃硫醇(2-methyl-3-tetrahydrofuranethiol,FEMA 号 3787)有顺、反两种异构体[9~12],调香中使用的是其异构体混合物。它是维生素 B_1 热降解的产物[13]。该香料为无色至浅黄色液体,稀释后具有肉香和烤肉香,是最重要的肉味香料之一[14],在香精中的贡献主要在头香和体香,可用于多种食用香精中,在肉类食品中的推荐用量为 0.5~2.0mg/kg,在快餐食品中的推荐用量为 5.0~15mg/kg[15]。

2-甲基-3-四氢呋喃硫醇可以通过下述四种方法合成:

1. 四氢呋喃路线

四氢呋喃和二氯磺酰在四氯化碳中反应制得 2,3-二氯四氢呋喃,产物为 100% 反式异构体,产率 60%~74%[16~18]。

$$\text{（四氢呋喃）} + SO_2Cl_2 \xrightarrow{CCl_4} \text{（2,3-二氯四氢呋喃）}$$

2,3-二氯四氢呋喃和甲基溴化镁在乙醚中反应制得 2-甲基-3-氯四氢呋喃[19~22]。

$$\text{（2,3-二氯四氢呋喃）} + CH_3MgBr \xrightarrow{(CH_3CH_2)_2O} \text{（2-甲基-3-氯四氢呋喃）}$$

2-甲基-3-氯四氢呋喃和硫代乙酸钾在 N,N-二甲基甲酰胺(DMF)中反应制得 2-甲基-3-乙酰硫基四氢呋喃[9,10]。

$$\text{（2-甲基-3-氯四氢呋喃）} + CH_3COSK \xrightarrow{DMF} \text{（2-甲基-3-乙酰硫基四氢呋喃）}$$

2-甲基-3-乙酰硫基四氢呋喃在碱性条件下水解、酸化生成 2-甲基-3-巯基四氢呋喃[9,10]。

$$\text{（2-甲基-3-乙酰硫基四氢呋喃）} + H_2O \xrightarrow{NaOH} \text{（2-甲基-3-巯基四氢呋喃）}$$

用乙基溴化镁代替甲基溴化镁,按上述合成路线可以合成 2-乙基-3-巯基四氢呋喃[23]。

2. 2-甲基四氢呋喃-3-酮路线

2-甲基四氢呋喃-3-酮用氢化铝锂或金属钠还原得 2-甲基四氢呋喃-3-醇[24,25]。

$$\text{（2-甲基四氢呋喃-3-酮）} + LiAlH_4 \longrightarrow \text{（2-甲基四氢呋喃-3-醇）}$$

2-甲基四氢呋喃-3-醇和氢溴酸反应,生成 2-甲基-3-溴四氢呋喃。

$$\text{（2-甲基四氢呋喃-3-醇）} + HBr \longrightarrow \text{（2-甲基-3-溴四氢呋喃）}$$

2-甲基-3-溴四氢呋喃和硫脲在醇中反应生成 S-(2-甲基四氢呋喃基)异硫脲氢溴酸盐,然后依次和碱、酸作用生成 2-甲基-3-巯基四氢呋喃。

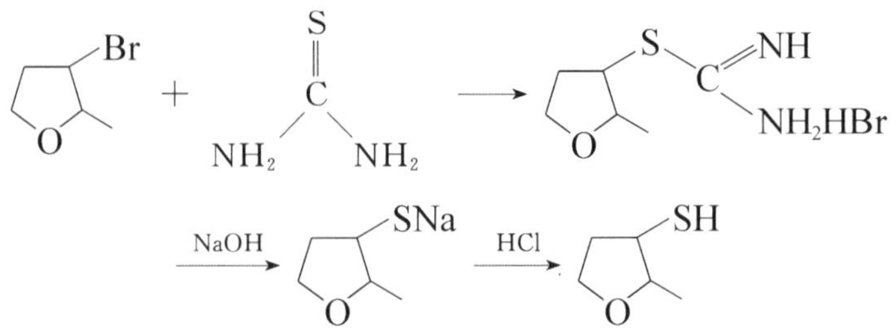

3. 2,3-二氢呋喃路线

2,3-二氢呋喃和溴在四氯化碳中反应,生成2,3-二溴四氢呋喃,产物为100%反式体[26]。

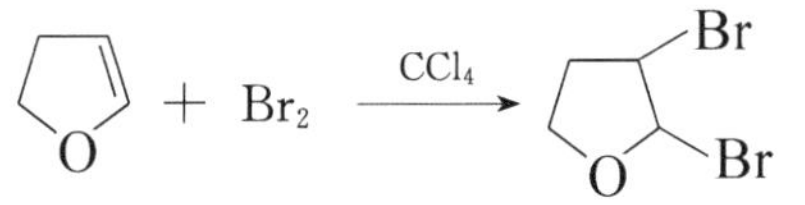

2,3-二溴四氢呋喃和甲基溴化镁在无水乙醚中反应生成2-甲基-3-溴四氢呋喃[21]。

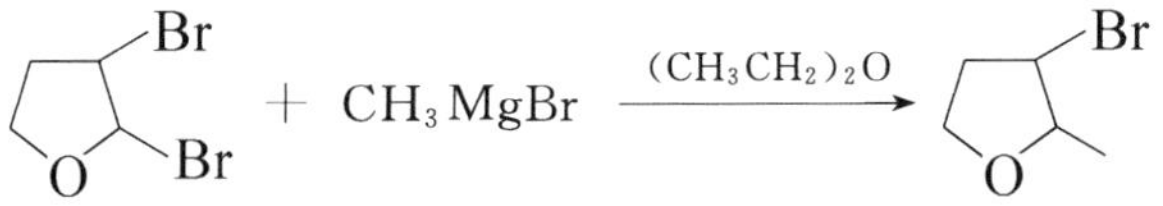

由2-甲基-3-溴四氢呋喃制取2-甲基-3-巯基四氢呋喃的方法同上。

4. 2-甲基-4,5-二氢呋喃路线

2-甲基-4,5-二氢呋喃和硫代乙酸加成得2-甲基-3-乙酰硫基四氢呋喃,2-甲基-3-乙酰硫基四氢呋喃在碱性条件下水解可制得2-甲基-3-巯基四氢呋喃[27]。

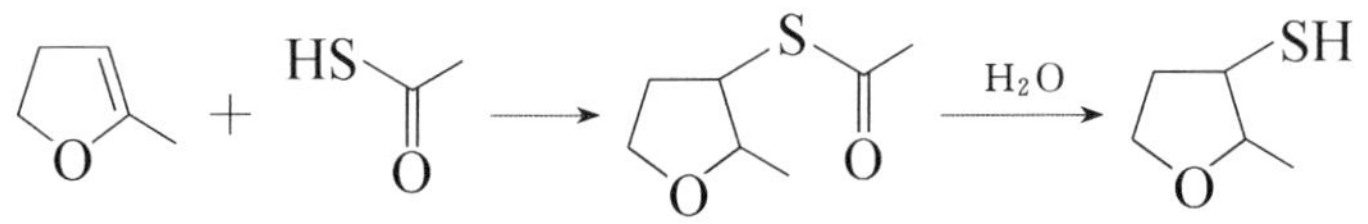

原料2-甲基-4,5-二氢呋喃的合成方法有以下几种:

(1) 由2-溴甲基四氢呋喃合成[28]

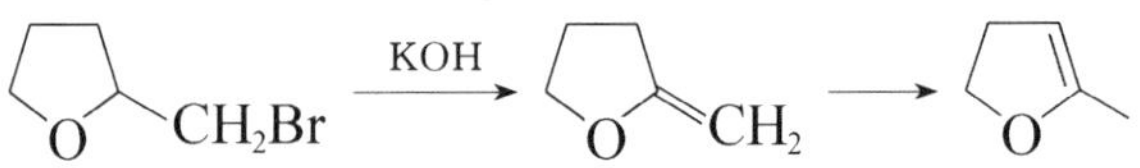

(2) 由2-甲基-3-氯四氢呋喃合成[29,30]

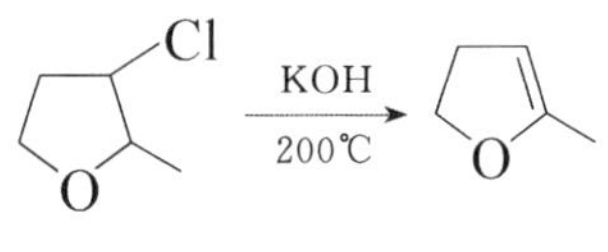

(3) 由乙酰丙醇合成[31～33]

$$CH_3\overset{O}{\overset{\|}{C}}CH_2CH_2CH_2OH \longrightarrow \underset{O}{\text{(2-methylfuran)}} + H_2O$$

其他一些 3-呋喃硫醇的化学结构和香味特征见表 3-1。

表 3-1 其他一些 3-呋喃硫醇的化学结构和香味特征

名称	化学结构式	香味特征
2-甲基-4-呋喃硫醇	(结构式：呋喃环，2位甲基，4位SH)	青香、肉香、药草香[34]
2-甲基-3-二氢呋喃硫醇	(结构式：二氢呋喃环，2位甲基，3位SH)	烤肉香[34]

第三节　糠硫醇和烯丙硫醇

一、糠硫醇

糠硫醇(furfuryl mercaptan,FEMA 号 2493)又称 2-呋喃基甲硫醇、咖啡醛,是香油、咖啡的主要香成分,在鸡肉、牛肉、猪肉等香成分中都有发现。糠硫醇为无色透明油状液体,具有咖啡、芝麻油、鸡蛋、肉、菜肴等香气,以及咖啡、芝麻、烤香、洋葱、大蒜等味道。可用于咖啡、芝麻、巧克力、洋葱、大蒜、牛肉、鸡肉、猪肉及烤香味等食品香精。

二糠基二硫醚在乙醇溶液中,用锌粉和乙酸还原,或用活性氧化铝进行还原可生成糠硫醇,糠基氯与硫脲缩合,然后再水解也可以生成糠硫醇。但目前最常用的方法是用糠醇与硫脲缩合,然后再水解制取[35]。

$$\text{呋喃-CH}_2\text{OH} + \underset{NH_2}{\overset{S}{\underset{\|}{C}}}NH_2 \xrightarrow{NaOH} \text{呋喃-CH}_2SNa \xrightarrow{HCl} \text{呋喃-CH}_2SH$$

二、烯丙硫醇

烯丙硫醇(allyl mercaptan,FEMA 号 2035)为无色液体,不溶于水,溶于乙醇等有机溶剂。沸点为 67～68℃,折射率 n_D^{20} 1.4765。该化合物存在于洋葱、大蒜中,具有强烈的洋葱、大蒜、韭葱香气和味道,可用于肉味、调味品、洋葱、大蒜等食品香精。

烯丙硫醇可以由烯丙醇和硫化氢在二氧化钍存在下,经加热缩合反应制取[1]。

$$\diagup\!\!\!\diagdown\!\!\text{OH} + H_2S \xrightarrow{ThO_2} \diagup\!\!\!\diagdown\!\!\text{SH} + H_2O$$

第四节　不含其他官能团的单硫醇类香料

一、苄硫醇

苄硫醇(benzyl mercaptan,FEMA 号 2147)为无色液体,具有强烈的洋葱、大蒜香气和味道,在咖啡、猪肉中有微量存在,主要用于咖啡、韭葱、洋葱、大蒜、西红柿、辣根、肉味等食品香精。

苄硫醇可以由氯化苄与硫氢化钾反应制取[1]。

$$\text{Ph-CH}_2\text{Cl} + \text{KSH} \xrightarrow{\triangle} \text{Ph-CH}_2\text{SH} + \text{KCl}$$

二、1-对-䓝烯-8-硫醇

1-对-䓝烯-8-硫醇(1-p-menthene-8-thiol,FEMA 号 3700)天然发现于葡萄柚汁中,是葡萄柚的特征香味成分。它为无色液体,具有葱蒜、圆柚、洋葱、大蒜、热带水果、木香等香气,以及圆柚、热带水果、芒果等味道。可用于圆柚、榴莲、葡萄、热带水果、芒果、黑醋栗等食品香精。

1-对-䓝烯-8-硫醇可以用柠檬烯为起始原料通过下述路线制备[36]:

也可以由 α-松油醇与氢溴酸、硫脲反应制备[1]。

第五节　β-羟基硫醇类香料

一、β-羟基硫醇的合成方法

1. 以 α,β-环氧化物为中间体或原料反应

(1) 同时合成 β-羟基硫醇和 β-巯基硫醇[37]

$$O=C=S + NH_3 \xrightarrow{室温} H_2N-\underset{\underset{O}{\|}}{C}-SH \cdot NH_3$$

$$H_2N-\underset{\underset{O}{\|}}{C}-SH \cdot NH_3 + \underset{R_2\ \ \ X\ \ \ R_4}{R_1\diagup\!\!\!\diagdown R_3}$$

$$\xrightarrow[醇(溶剂)]{15\sim 19℃} \underset{\underset{HS\ \ XH}{R_2\ \ \ R_4}}{R_1\ \ \ R_3} + H_2N-\underset{\underset{O}{\|}}{C}-NH_2$$

式中:X 为 O 或 S;$R_1\sim R_4$ 为取代基,可以是 H、烷基、芳基等。氨可以用伯胺代替,如甲胺、乙胺、丙胺等;溶剂为低相对分子质量的脂肪醇,如甲醇、乙醇、丙醇等。

合成 2-巯基丙醇的产率为 53.8%。

(2) 通过四步反应合成 2-甲基-2-巯基戊醇[38]

$$\underset{H}{\overset{C_2H_5}{\diagdown}}C=C\underset{CHO}{\overset{CH_3}{\diagup}} \xrightarrow{LiAlH_4} \xrightarrow[CH_2Cl_2]{间氯过氧苯乙酸} \underset{H}{\overset{C_2H_5}{\diagdown}}\underset{O}{\triangle}\underset{CH_2OH}{\overset{CH_3}{\diagup}}$$

$$\xrightarrow[四异丙氧基钛,THF]{(NH_2)_2CS} \xrightarrow[C_2H_5OC_2H_5]{NaHCO_3(饱和水溶液)} \underset{H}{\overset{C_2H_5}{\diagdown}}\underset{S}{\triangle}\underset{CH_2OH}{\overset{CH_3}{\diagup}}$$

$$\xrightarrow{NaAlH_2(CH_3OCH_2CH_2O)_2} \xrightarrow{H_3O^+} \underset{SH}{\overset{|}{\underset{|}{\diagdown}}}\!\!\!\!\diagdown\!\!\!OH$$

(3) α,β-环氧化物和硫化氢反应

硫化氢与各种氧杂环丙烷进行开环反应是制取 β-羟基硫醇的常用方法之一[39~42]:

$$\underset{R_2\ \ \ O\ \ \ R_4}{R_1\diagup\!\!\!\diagdown R_3} + H_2S \longrightarrow \underset{\underset{HO\ \ SH}{R_2\ \ \ R_4}}{R_1\ \ \ R_3}$$

此反应产率仅有 50%~60%,因为此反应过程中产物巯基醇会和 α,β 环氧化物进一步反应生成二醇硫醚。为减少此种化合物的生成,可预先向反应物中加入催化剂量的二醇硫醚或巯基醇,同时通入过量的硫化氢等[39~42]提高产率。

(4) α,β-环氧化物与 $NaBH_2S_3$ 反应

$NaBH_2S_3$ 与环氧化物发生开环反应,然后水解,再用 $LiAlH_4$ 还原可制得相应的 β-羟基硫醇,产率为 83%~91%[43,44]。

2. 烯醇与硫化氢加成

3-甲基-2-丁烯醇与硫化氢在紫外线照射下发生自由基加成反应生成 3-甲基-2-巯基醇,产率为 97%[45]。

3. 硫脲与卤代醇反应

硫脲与卤代醇反应生成卤化锍盐,然后用碱水解,再酸化可生成 β-羟基硫醇[46]。

4. 烷基黄原酸盐与 α-卤代醇或 α-卤代羰基化合物反应

甲基黄原酸钾与 α-卤代醇或 α-卤代羰基化合物反应生成相应的甲基黄原酸酯,再用 LiAlH₄ 还原,可制取相应的 β-羟基硫醇[46]。

5. 羰基化合物还原

1) 酮的 α-H 卤代生成 α-卤代酮,再与硫氢化物反应生成 α-巯基酮,然后用氢化铝锂或硼氢化钠还原可得 β-羟基硫醇[47,48]。

2) 苄硫醇与 α-卤代羰基化合物反应生成 α-苄硫基羰基化合物,然后进行 α-活泼氢的烷基取代,再还原羰基,最后脱去苄基生成相应取代 β-羟基硫醇[49]。

$$\xrightarrow{RI} \text{Ph} \underset{}{\overset{R}{\text{S}}}\underset{O}{\overset{}{\text{C}}}\text{R}' \longrightarrow \text{Ph}\underset{}{\overset{R}{\text{S}}}\underset{OH}{\overset{}{\text{C}}}\text{R}' \longrightarrow \text{HS}\underset{OH}{\overset{R}{\text{C}}}\text{R}'$$

6. β-羟基二硫醚还原

β-羟基二硫醚用四氢锂铝还原生成β-羟基硫醇[50]。

$$\underset{R_2\ R_1}{\overset{OH}{\text{C}}}\text{S}-\text{S}\underset{R_1}{\overset{OH}{\text{C}}}\text{R}_2 \xrightarrow{LiAlH_4} \underset{R_2\ R_1}{\overset{OH}{\text{C}}}\text{SH}$$

二、β-羟基硫醇类香料化合物

β-羟基硫醇类香料中允许作为食品香料使用的有 3-巯基-2-丁醇（FEMA 号 3502）及 2-甲基-2-巯基-1-戊醇（FEMA 号 3995）。

3-巯基-2-丁醇，俗称"935"，是无色透明液体，具有烤肉、葱蒜、洋葱、大蒜等香气，以及肉、菜肴、洋葱、大蒜等味道，可用于鸡肉、烤牛肉、猪肉、洋葱、大蒜、调味品及各种菜肴香精。

3-巯基-2-丁醇可以通过 2,3-环氧丁烷与硫化氢反应制备[51]。

$$\triangle\text{O} + H_2S \longrightarrow \underset{SH}{\overset{OH}{\text{CHCH}}}$$

β-羟基硫醇类香料也可以通过相应的β-羰基硫醇还原制备[52]。

$$R_1\underset{SH}{\overset{O}{\text{C}}}\text{R}_2 \xrightarrow{NaBH_4} R_1\underset{SH}{\overset{OH}{\text{CH}}}\text{R}_2$$

一些β-羟基硫醇的化学结构和香味特征见表 3-2。

表 3-2 一些β-羟基硫醇的化学结构和香味特征

名称	化学结构式	香味特征
3-巯基-4-庚醇	(结构式：OH/SH)	硫化物样、青香、圆柚香、肉香[52]
4-巯基-5-壬醇	(结构式：OH/SH)	硫化物样、果香、圆柚香、橡胶味、苦味[52]

第六节 β-烷氧基硫醇类香料

β-烷氧基硫醇的一般合成方法如下:

(1) β-烷(芳)氧基卤代烃的亲核取代反应

以 β-烷氧基卤乙烷为原料通过下列反应可以合成 β-烷氧基乙硫醇[53]。

$$RO\text{-}CH_2CH_2\text{-}X + KSH \longrightarrow RO\text{-}CH_2CH_2\text{-}SH$$

式中:R 为甲基、乙基、丙基、丁基。

β-芳氧基卤乙烷与硫氢化钠直接进行亲核取代反应,生成相应的 β-芳氧基乙硫醇[54]。

$$ArO\text{-}CH_2CH_2\text{-}X + NaSH \longrightarrow ArO\text{-}CH_2CH_2\text{-}SH$$

(2) 1,3-氧硫杂环戊烷的开环反应

醛或酮与巯基乙醇缩合生成 1,3-氧硫杂环戊烷,再用金属和液氨还原开环制得 β-烷氧基乙硫醇[55,56]。

$$\underset{S}{\overset{O}{\diagdown}}\!\!\!\diagup\!\!\!\underset{R_2}{\overset{R_1}{\diagdown}} \xrightarrow{M, 液氨} R_1\underset{R_2}{\overset{}{\diagdown}}CH\text{-}O\text{-}CH_2CH_2\text{-}SH$$

式中:M 为 K、Na、Li、Ca;R_1、R_2 为 H 或烃基。

(3) 烷氧基甲基硫杂环丙烷还原

3-烷氧基-1,2-环硫丙烷用 $LiAlH_4$ 还原开环制得 3-烷氧基-2-丙硫醇[57]。

$$RO\text{-}CH_2\text{-}\underset{S}{\overset{}{\triangle}} \xrightarrow{LiAlH_4} RO\text{-}CH_2\text{-}CH(SH)\text{-}CH_3$$

式中:R 为 CH_3、C_2H_5、$n\text{-}C_3H_7$、$n\text{-}C_4H_9$、C_6H_5。

(4) β-烷氧基卤代烃与硫脲反应

β-烷氧基卤代烃与硫脲反应可以合成 β-烷氧基硫醇,如 1-甲氧基-2-丙硫醇和 1-乙氧基-2-丙硫醇的合成[57]。

$$RO\text{-}CH_2\text{-}CH(Br)\text{-}CH_3 \xrightarrow{NH_2CSNH_2} \xrightarrow[H_2O]{NaOH} RO\text{-}CH_2\text{-}CH(SH)\text{-}CH_3$$

式中:R 为 CH_3 或 C_2H_5。

(5) 醇与硫杂环丙烷反应

在催化剂 $BF_3\text{-}HAc$ 的作用下,醇与硫杂环丙烷发生开环反应可生成相应的 β-烷氧基硫醇[58,59]。

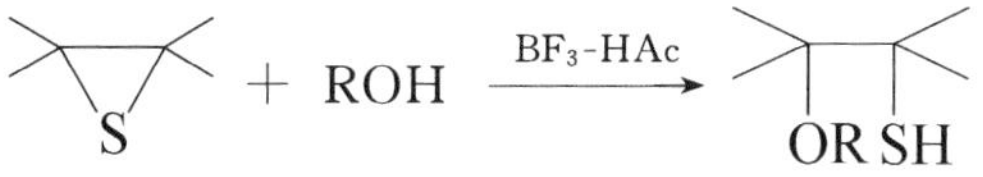

$$\underset{S}{\triangle} + ROH \xrightarrow{BF_3-HAc} \underset{OR\ SH}{\diagdown\diagup}$$

第七节 γ-羟基硫醇类香料

已经允许作为食品香料使用的 γ-羟基硫醇类化合物有 2-甲基-3-巯基-1-丁醇（FEMA 号 3993）、3-甲基-3-巯基-1-丁醇（FEMA 号 3854）、2-甲基-3-巯基-1-戊醇（FEMA 号 3996）、3-巯基-1-己醇（FEMA 号 3850）等。

γ-羟基硫醇类化合物可以通过下面的方法制备：

(1) α-烯酮与硫化氢反应[60~62]

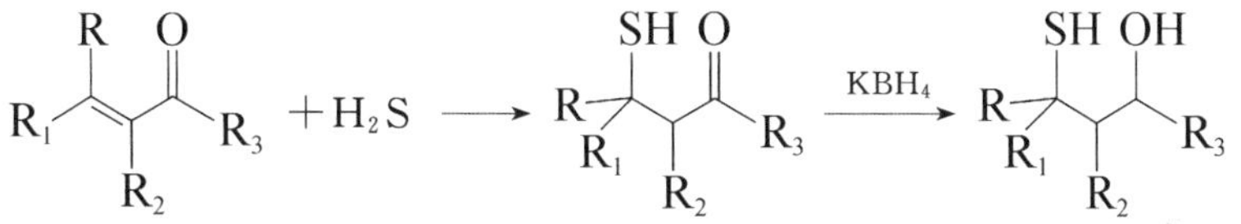

(2) α-烯酮与硫代乙酸反应[61,63]

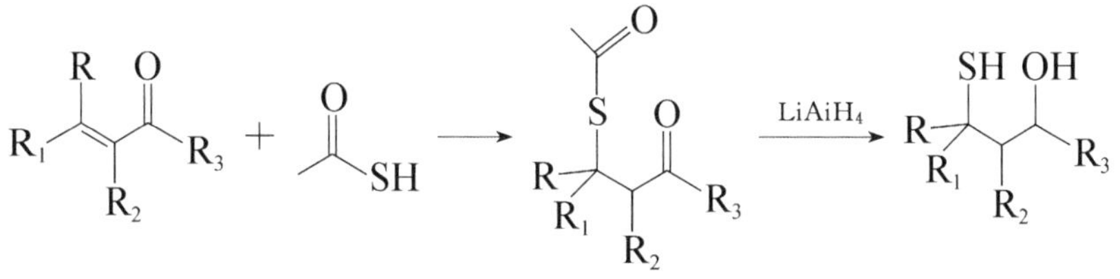

一些 γ-羟基硫醇的化学结构和香味特征见表 3-3。

表 3-3　一些 γ-羟基硫醇的化学结构和香味特征

名称	化学结构式	香味特征
3-巯基-1-丙醇	 SH OH	土豆香、肉汤香[63]
3-巯基-1-丁醇	 SH OH	奶酪香、洋葱香、甘蓝香、刺激性的香气[63]
2-甲基-3-巯基-1-丙醇	 SH OH	韭菜香、洋葱香[63]
3-甲基-3-巯基-1-丁醇	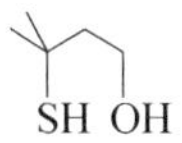 SH OH	煮洋葱香、酱油香、山萝卜香[63]
2-甲基-3-巯基-1-丁醇	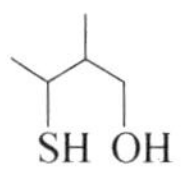 SH OH	韭菜香、洋葱香、胡椒香[63]

续表

名称	化学结构式	香味特征
3-巯基-1-戊醇		肉汤香、青香[63]
2-乙基-3-巯基-1-丙醇		洋葱香[63]
2-甲基-3-巯基-1-戊醇		塑料燃烧、汽油、柑橘类水果香、青香、醋香、酱油香[63]
3-巯基-1-己醇		葡萄香、鸡蛋果样、大黄、酸柠檬香[63]
3-巯基-1-庚醇		柑橘类水果香、醋香、酱油香、胡萝卜香[63]
2-丁基-3-巯基-1-丙醇		塑料、刺激性香气[63]
3-巯基-1-辛醇		大黄、胡萝卜香、青香[63]
3-巯基-1-壬醇		胡萝卜香、青香[63]
4-巯基-2-戊醇		金雀花香、黑醋栗香、生洋葱香[64]
4-巯基-4-甲基-2-戊醇		金雀花香、黑醋栗香、溶剂、清新、甜的香气[64]
3-甲基-3-巯基-1-己醇（FEMA 号 3854）		药草香、壤香、青香、热带水果香、圆柚、鸡蛋、果香、黑醋栗香、洋葱香[65]
5-巯基-3-己醇		汗味、肉汤香、柑橘类水果香、煮牛奶香[64]
4-巯基-3-甲基-2-戊醇		洋葱香、韭菜香、汗味、肉汤香[64]
4-巯基-5-甲基-2-己醇		大黄、柠檬香、辛香、胡椒香、肉香[64]

第八节 β-羰基硫醇类香料

β-羰基硫醇类香料在含硫香料中占有重要的地位,这类香料中的巯基丙酮(FEMA号3856)、3-巯基-2-丁酮(FEMA号3298)、3-巯基-2-戊酮(FEMA号3300)都是允许使用的食品香料,此类香料一般可以通过下述路线合成[2,52]:

$$R_1\text{-CO-CHX-}R_2 + \text{NaSH} \longrightarrow R_1\text{-CO-CH(SH)-}R_2$$

式中:X 为 Cl 或 Br;所需卤代酮可以由酮与 SO_2Cl_2 [2,52] 或 Br_2 [3] 反应制备。

β-羰基硫醇类香料常具有令人愉快的肉香味[66],一些 β-羰基硫醇的化学结构和香味特征见表 3-4。

表 3-4 一些 β-羰基硫醇的化学结构和香味特征

名称	化学结构式	香味特征
巯基丙酮		肉香
3-巯基-2-丁酮		肉香[66]
3-巯基-3-甲基-2-丁酮		肉香[66]
2-巯基-2-甲基-3-戊酮		肉香[66]
3-巯基-2-戊酮		肉香
3-巯基-3-甲基-2-戊酮		肉香[66]
2-巯基-3-戊酮		醚香、甜的、硫化物样、肉香、葱蒜样香气;硫化物样、烤香、圆柚样、肉香香味[2,52]

名称	化学结构式	香味特征
4-巯基-4-甲基-3-己酮		肉香[66]
3-巯基-4-庚酮		醚香、果香[2]
2,6-二甲基-3-巯基-4-庚酮		清香、果香、圆柚香、柑橘香、长叶薄荷酮硫醇香；圆柚香、柠檬香、柑橘香、黑醋栗香味[2,52]
4-巯基-5-壬酮		硫化物、果香、橡胶样、圆柚香气；果香、硫化物、柑橘和圆柚香味[52]
2-巯基环十二酮		花香、蔬菜清香、薄荷香、樟脑香、硫化物香气；蔬菜清香、薄荷香、凉香、硫化物香味[52]
α-巯基苯乙酮		肉香、明显的鸡肉香气[2]
R 取代 α-巯基苯乙酮 R 为 1～4 碳烷基、烷氧基、氢、羟基、乙酰基、苯乙酰基、苄基		肉香[67]
4-巯基四氢呋喃-3-酮		青香、肉香、肉汁香[34]
5-甲基-4-巯基四氢呋喃-3-酮		肉香、肉汁香[34]

第九节 γ-羰基硫醇类香料

4-巯基-2-丁酮(FEMA 号 3357)、2-甲基-3-巯基戊醛(FEMA 号 3994)、4-甲基-

4-巯基-2-戊酮(FEMA 号 3997)是允许使用的 γ-羰基硫醇类香料,这类香料可以用下面几种方法合成[60,61]：

(1) α-烯酮与硫化氢反应[60,61]

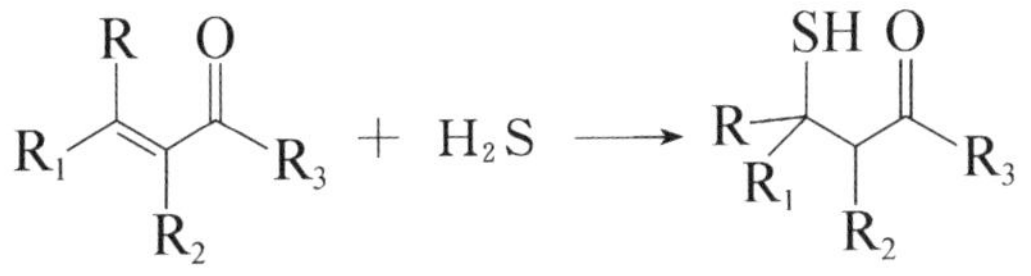

(2) α-烯酮与硫代乙酸反应[61,63]

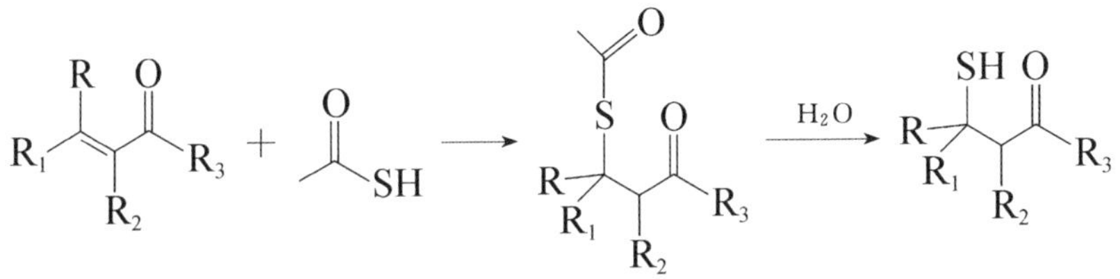

γ-羰基硫醇类香料中有必要重点介绍的是硫代薄荷酮,又称对䓝-8-硫醇-3-酮(FEMA 号 3177)。该香料市售品为多种异构体混合物,具有薄荷、青香、水果、热带水果、布枯香气和味道,常用于调配葡萄、薄荷、覆盆子、草莓、桃子、热带水果香精。

硫代薄荷酮有四个异构体,其香气特征分别为:2R-trans 型具有强烈的带霉味的硫化物香韵;2S-trans 型具有强烈而持续的青的黑加仑香韵;2S-cis 型具有柔和的、甜的、果香韵;2R-cis 型具有强烈的甜香,稍微有一点刺激性。

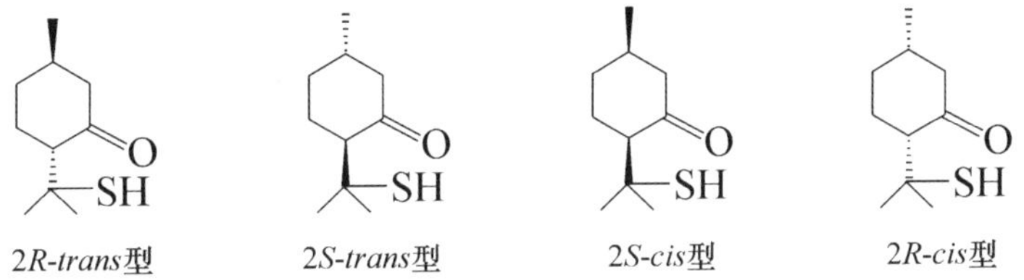

2R-trans型   2S-trans型   2S-cis型   2R-cis型

硫代薄荷酮可以用胡薄荷酮为原料制备[1]。

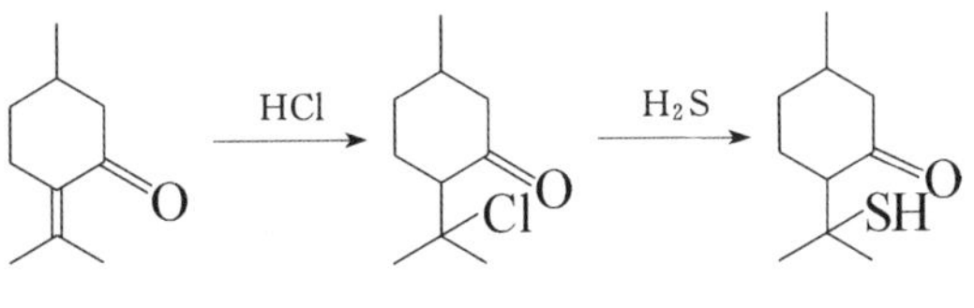

一些 γ-羰基硫醇的化学结构和香味特征见表 3-5。

表 3-5　一些 γ-羰基硫醇的化学结构和香味特征

名称	化学结构式	香味特征
3-巯基丁醛	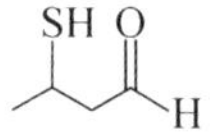	肉汤香、奶酪香、辛辣香[60]
2-甲基-3-巯基丁醛	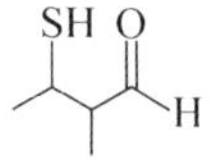	肉汤香、洋葱香、肉香、奶油香[60]
3-甲基-3-巯基丁醛	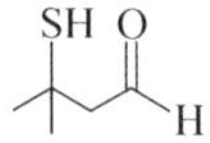	肉汤香、奶酪香、刺激的香气[60]
4-巯基-2-丁酮	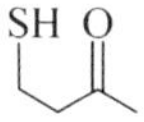	土豆香[63]
4-巯基-2-戊酮	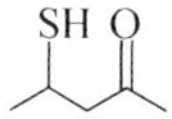	青香、土豆香、黑醋栗香[60,64]
3-甲基-4-巯基-2-戊酮	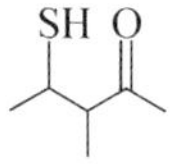	汗味、煮牛奶香[60,64]
4-甲基-4-巯基-2-戊酮	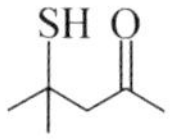	黑醋栗香、蔬菜香、柑橘香、金雀花香[60,64]
1-巯基-3-戊酮	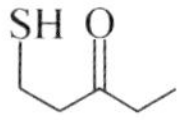	奶酪香、臭鼬、溶剂、刺激的香气[63]
5-巯基-3-己酮	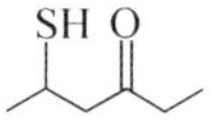	清新的、焦臭的香气[63,64]
5-甲基-4-巯基-2-己酮	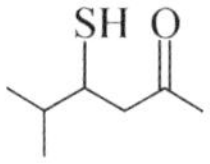	水果香、甜香、硫化物样、甘蓝香[63,64]
4-巯基-2-壬酮	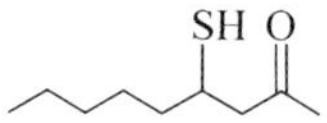	大黄、印度大麻、柠檬香、辛香[63]

第十节　α,β-二硫醇类香料

一、α,β-二硫醇的一般合成方法

1. 炔烃与硫化氢加成

在常温下,密闭容器中用 X 射线照射硫化氢与炔烃发生自由基加成反应,生成烯硫醇、α,β-二硫醇及少量聚合物[68]。

$$R\!\equiv\!R' + H_2S \longrightarrow \underset{SH}{R\!\!-\!\!\overset{R'}{=}} + \underset{R\!\!-\!\!\underset{SH}{|}\!\!-\!\!\underset{}{\overset{SH}{|}}\!\!-\!\!R'}{}$$

2. 共轭双烯与硫化氢的加成

在水、自由基引发剂(如紫外线、二叔丁基过氧化物等)和金属(如 Fe、Co、Ni 等)的存在下,用 2,5-二甲基-2,4-己二烯与过量的硫化氢发生加成反应,生成相应的硫醇和 α,β-二硫醇[69,70]。

3. 烯烃与硫为原料的反应

在惰性溶剂存在下,将烯烃与硫粉在高温(175℃左右)和高压(大于 7atm①)下反应后,在 75~300℃、加压(大于 10atm)下,用 H_2 和金属多硫化物(如 CoS_3)还原,可以制得 α,β-二硫醇[71,72]。

4. α,β-环硫化物的开环反应

α,β-环硫化物与硫氢化钾或硫氢化钠在催化剂作用下发生开环反应,生成 α,β-二硫醇[73,44]。

已经允许作为食品香料使用的 α,β-二硫醇类香料有 1,2-乙二硫醇(FEMA 号 3484)、1,2-丙二硫醇(FEMA 号 3520)、1,2-丁二硫醇(FEMA 号 3528)及 2,3-丁二硫醇(FEMA 号 3477)。

二、1,2-乙二硫醇

1,2-乙二硫醇(1,2-ethanedithiol)是结构最简单的 α,β-二硫醇类香料,为无色透明液体,发现于清炖牛肉香成分中,具有葱、蒜、肉、烤肉等香味[74],可用于调配葱、蒜、肉等食品香精。

1,2-乙二硫醇合成方法很多,常见的有以下六种方法:

---

① 1atm=1.013 25×10^5Pa,下同。

(1) 1,2-二卤乙烷与硫氢化钠反应

$$\begin{matrix} X \\ X \end{matrix} + NaSH \longrightarrow \begin{matrix} SH \\ SH \end{matrix} + NaX$$

式中:X 为 Br、Cl。常用的二卤代烷是 1,2-二溴乙烷[75]和 1,2-二氯乙烷[76]。

(2) 1,2-乙二醇与五硫化二磷反应[77]

$$\begin{matrix} OH \\ OH \end{matrix} + P_2S_5 \longrightarrow \begin{matrix} SH \\ SH \end{matrix}$$

(3) 环硫乙烷与硫化氢反应[78]

$$\triangle_S + H_2S \longrightarrow \begin{matrix} SH \\ SH \end{matrix}$$

(4) 1,2-二溴乙烷与硫脲反应[79,80]

$$\begin{matrix} CH_2Br \\ CH_2Br \end{matrix} + S=C\begin{matrix} NH_2 \\ NH_2 \end{matrix} \longrightarrow \begin{matrix} CH_2-S-C(NH)-NH_2 \cdot HBr \\ CH_2-S-C(NH)-NH_2 \cdot HBr \end{matrix} \xrightarrow[2)\ HCl]{1)\ NaOH} \begin{matrix} CH_2SH \\ CH_2SH \end{matrix}$$

(5) 二丙胺与硫化氢反应,再与 1,2-二溴乙烷反应[81]

$$(C_3H_7)_2NH + H_2S \longrightarrow (C_3H_7)_2NH_2SH \xrightarrow{BrCH_2CH_2Br} \begin{matrix} SH \\ SH \end{matrix}$$

(6) 硫氢化铵与乙酸(2-巯基乙醇)酯反应[82]

$$NH_4SH + HSCH_2CH_2OAc \longrightarrow \begin{matrix} SH \\ SH \end{matrix}$$

三、2,3-丁二硫醇

2,3-丁二硫醇(2,3-butanedithiol)是最重要的脂肪族邻二硫醇类香料,商品名"865",为无色至浅黄色液体,其香气特征为硫化物样、肉香、大蒜和咖啡香韵;味觉特征为硫化物样、脂肪样、烤牛肉、鸡蛋、意大利腊肠和猪肉香韵[83,84]。可用于烤牛肉、烤猪肉、炖牛肉、炖猪肉、意大利腊肠、鸡蛋、咖啡、大蒜等多种食用香精,在肉类食品和焙烤食品中的建议用量为 0.2mg/kg[85]。

2,3-丁二硫醇可以用 2,3-环氧丁烷为原料,与硫氰酸钾反应生成 2,3-环硫丁烷,然后再与硫氢化钾的乙醇溶液反应合成[84]。

$$\triangle_O \xrightarrow{KSCN} \triangle_S \xrightarrow{KSH} \begin{matrix} SH \\ SH \end{matrix}$$

也可以采用硫脲和2,3-环氧丁烷制备2,3-环硫丁烷,进而与硫氢化钠加成的方法合成[86]。

$$\text{环氧丁烷} \xrightarrow{(NH_2)_2CS} \text{环硫丁烷} \xrightarrow{NaSH} \text{2,3-丁二硫醇}$$

第十一节 α,ω-二硫醇类香料

1,3-丙二硫醇(FEMA 号 3588)、1,6-己二硫醇(FEMA 号 3495)、1,8-辛二硫醇(FEMA 号 3514)、1,9-壬二硫醇(FEMA 号 3513)等是典型的 α,ω-二硫醇类香料。1,2-乙二硫醇也是 α,ω-二硫醇类香料。这类香料可以用下面的结构通式表示:

$$HSCH_2(CH_2)_nCH_2SH$$

式中:n 为 0 或正整数。

α,ω-二硫醇类香料一般可以采用相应的二卤代烷与硫脲反应制备[87~89]。

$$\begin{array}{c} CH_2X \\ (CH_2)_n \\ CH_2X \end{array} + S{=}C\begin{array}{c}NH_2\\NH_2\end{array} \longrightarrow \begin{array}{c}CH_2S{-}C({=}NH){-}NH_2 \cdot HX \\ (CH_2)_n \\ CH_2S{-}C({=}NH){-}NH_2 \cdot HX\end{array} \xrightarrow[2)\,HCl]{1)\,NaOH} \begin{array}{c}CH_2SH \\ (CH_2)_n \\ CH_2SH\end{array}$$

式中:X 为 Br 或 Cl;n 为 0 或正整数;常用的二卤代烷是二溴代烷和二氯代烷。

一些 α,ω-二硫醇类香料的化学结构和香味特征见表 3-6。

表 3-6 一些 α,ω-二硫醇类香料的化学结构和香味特征

名称	化学结构	香味特征
1,3-丙二硫醇	$HSCH_2CH_2CH_2SH$	稀释后具有蛋黄香味
1,6-己二硫醇	$HSCH_2(CH_2)_4CH_2SH$	鸡肉特征香[89]
1,8-辛二硫醇	$HSCH_2(CH_2)_6CH_2SH$	洋葱香、大蒜香
1,9-壬二硫醇	$HSCH_2(CH_2)_7CH_2SH$	洋葱香、大蒜香、肉香[14]

第十二节　硫酚类香料

目前允许使用的硫酚类香料主要有苯硫酚(FEMA 号 3616)、2-甲基苯硫酚(FEMA 号 3240)、2,6-二甲基苯硫酚(FEMA 号 3666)、2-乙基苯硫酚(FEMA 号 3345)、2-萘硫酚(FEMA 号 3314)等。

一些硫酚类香料的化学结构和香味特征见表 3-7。

表 3-7　一些硫酚类香料的化学结构和香味特征

名称	化学结构式	香味特征
苯硫酚	C₆H₅—SH	肉香、焦香、酚样香气和味道
2-甲基苯硫酚		肉香、肉汤香、洋葱香、大蒜香、鸡蛋香气;肉香、洋葱香、大蒜味道[90]
4-甲基苯硫酚		焦香、青香、脂肪香[90]
3,4-二甲基苯硫酚		焦香[90]
2,6-二甲基苯硫酚		肉香[90]
2-乙基苯硫酚		烟熏、肉香香气及酚样、烟熏味道
2-萘硫酚		油腻、烤香、汤料香、蘑菇香、肉香香气;肉香、烤香、鸡肉香、鸡蛋香、坚果香味道

第十三节　含六元芳香杂环的硫醇类香料

一、2-吡啶甲硫醇

2-吡啶甲硫醇(2-pyridinemethanethiol, FEMA 号 3232)具有烤香、肉香、牛肉香、菜肴、猪肉香气以及肉香、烤香、菜肴、脂肪、牛肉、爆玉米花味道,可用于猪肉、牛肉、鸡肉、菜肴、蔬菜、烟熏等食品香精[91]。

2-吡啶甲硫醇可以由 2-吡啶甲醛为原料制备[92],也可以由 2-吡啶基甲基氯与硫氢化钠反应制备。

$$\text{Py-CH}_2\text{Cl} + \text{NaSH} \longrightarrow \text{Py-CH}_2\text{SH} + \text{NaCl}$$

二、2-吡嗪甲硫醇

2-吡嗪甲硫醇(2-mercaptomethylpyrazine,FEMA 号 3299)又称 2-巯基甲基吡嗪,无色至淡黄色液体,具有烤香香气及烤肉样味道,主要用来调配肉味、坚果等食品香精[93]。

2-吡嗪甲硫醇可以由氯甲基吡嗪与硫氢化钠反应制备。

$$\text{Pyrazine-CH}_2\text{Cl} + \text{NaSH} \longrightarrow \text{Pyrazine-CH}_2\text{SH} + \text{NaCl}$$

三、吡嗪乙硫醇

吡嗪乙硫醇(pyrazineethanethiol,FEMA 号 3230)又称 2-吡嗪基乙硫醇,具有肉香、烤香、烤牛肉、烤鸡肉香气以及肉香、清炖、鸡肉味道,可用于牛肉、猪肉、鸡肉、咖啡、烤香等食品香精[91]。

参 考 文 献

[1] 孙宝国,何坚. 香料化学与工艺学. 第二版. 北京:化学工业出版社,2004. 470~489
[2] 高振永,柯友之,王亨权等. 合成食品香味增效剂 α-羰基硫醇和 α,β-二羰基硫醚的研究. 北京轻工业学院学报,1985,3(1):46~56
[3] Minabe M, Suzuki K. Fluorene derivatives XXXI. J. Org. Chem., 1975, 40(9):1297
[4] Cossar B C, Fournier J O, Fields D L et al. Preparation of thiols. J. Org. Chem., 1962, 27(1):93~94
[5] Cunneen J I. The Addition of thio-compounds to olefins. J. Chem. Soc., 1947, 1:134~147
[6] 孙宝国,田红玉,郑福平等. 肉香味含硫香料及其合成方法//中国科学技术协会. 绿色高新精细化工技术. 北京:化学工业出版社,2004. 180~191
[7] 孙宝国,钟香驹,梁梦兰. 2-甲基-2,5-二甲基-2,5-二氢呋喃合成研究. 精细化工,1991,(5):24~25
[8] 孙宝国,钟香驹,梁梦兰. 2-甲基-2,5-二甲氧基-2,5-二氢呋喃水解平衡的研究. 北京轻工业学院学报,1992,10(1):48~50
[9] Ouweland V D, Godefridus A M, Peer H G. Flavoring substances:US, 4134901. 1979. 1.16
[10] Ouweland V D, Peer H G. Flavoring substances:US, 4080367. 1978.3.21
[11] Ouweland V D, Godefridus A M, Peer H G. Certain lower alkyl 4,5-dihydrothiophene-3-thiols:US, 4020170. 1977.4.26
[12] Van Den Bosch S, Kettenes D K, De Roos K B et al. Sulfur-containing flavoring agents:Can, 1051441. 1979.3.27

[13] Hartman G J, Carlin J T, Scheide J D et al. Volatile products formed from the thermal degradation of thiamin a high and low moisture. J. Agric. Food Chem., 1984, 32(5): 1015~1018

[14] Rowe D J. Aroma chemicals for savory flavors. Perfumer & Flavorist, 1998, 23(4): 9~16

[15] Smith R L, Ford R A. Recent progress in the consideration of flavoring ingredients under the food additives amendment, 16. GRAS Substances. Food Technology, 1993, 47(6): 104~117

[16] 徐晶,孙宝国,郑福平等. 2,3-二氯四氢呋喃的简便合成方法//第三次全国精细化工青年科技学术交流会委员会. 第三次全国精细化工青年科技学术交流会论文集. 大连：大连出版社,1998. 124~125

[17] Crombie L, Wyvill R D. A halo ether synthesis of olefinic alcohols: stereochemistry and conformation of 2-substituted 3-halotetrahydropyran and furan precursors. J. Chem. Soc. (Perkin Trans. 1), 1985, (9): 1971~1978

[18] Nersation A. An improved synthesis of 2,3-dichlorotetrahydrofuran. Ind. Eng. Chem., Prod. Res. Develop., 1963, 2: 138~140

[19] Crombie L, Harper S H. Stereochemical studies of olefinic compounds(II). Ring scission of 2-(1-haloalkyl) tetrahydrofurans and 3-halo-2-alkyltetra- hydropyrans as a route to 4-alken-1-ols of known configuration and as a method of chain extension by five methylene groups. J. Chem. Soc., 1950, 1708~1714

[20] Crombie L, Gold J, Harper S H et al. Stereochemical studies of olefinic compounds(V). Ring fission of 3-chlorotetrahydrofurans and pyrans. J. Chem. Soc., 1956, 136~142

[21] 尤田耙,张美芳,叶明福. 3-己烯-1-醇及其酯类的合成和香气研究. 中国科学技术大学学报,1991, 21(3): 7~11

[22] Ansell M F, Brown S S. Synthesis of olefinic acids using tetrahydro-β-halo-furan and pyran derives as intermediates. J. Chem. Soc., 1957, 1788~1795

[23] 徐晶,孙宝国. 2-乙基-3-乙酰硫基四氢呋喃的合成研究. 精细化工,1999,16(增):368~370

[24] Coxon J M, Hartshorn M P. Acetate participation in acyclic epoxide systems. Acid catalyzed rearrangement of trans- and cis-acetoxy-3,4- epoxypentanes, -4,5-epoxyhexanes, and -5,8-epoxyheptanes. J. Org. Chem., 1974, 39(8): 1142~1148

[25] Coxon J M, Hartshorn M P. Hydroxyl participation in acyclic epoxide systems. Acid catalyzed rearrangement of trans- and cis-acetoxy-3,4- epoxypentanes, -4,5-epoxyhexanes, and -5,8-epoxyheptanes. Aust. J. Chem., 1973, 26(1): 2521

[26] 赵晶. 4,5-二氢-2-甲基呋喃和烷基呋喃基硫醚的合成研究[硕士学位论文]. 北京：北京工商大学, 2004. 4~5

[27] Bordwell F G et al. Free radical addition of thiolacetic acid to some cyclic olefins. J. Am. Chem. Soc., 1957, 79:3493

- [28] Paul R, Tchelitcheff S. 2-Methylenetetrahydrofuran. Application to the preparation of branched ethylenic alcohols. Synthesis of sylvan. Bull. Soc. Chim. France., 1950: 520~526
- [29] James T L, Wellington C A. Thermal decomposition of 2-methyl-2,5-dihydrofuran. J. Chem. Soc., 1968, (10): 2398
- [30] 孙宝国. 肉香味含硫香料合成及其分子结构特征研究 [博士学位论文]. 北京: 清华大学, 2002. 5~6
- [31] Londergan T E, Hause N L, Schmitz W R. A New synthesis of the thiazole fragment of Vitamin B_1. J. Am. Chem. Soc., 1953, 75: 4456~4458
- [32] Schniepp L E, Geller H H, Korff R W. The preparation of acetopropyl alcohol and 1,4-pentanediol from methylfuran. J. Am. Chem. Soc., 1947, 69: 672~674
- [33] Swadesh S, Smith S, Dunlop A P. Mechanism of hydrogenation of 2-methylfuran. J. Org. Chem., 1951, 16: 476~479
- [34] Godefridus A M, Ouweland V D, Peer H G. Components contributing to beef flavor. Volatile compounds produced by the reaction of 4-Hydroxy-5-methyl-3(2H)-furanone and tts thio analog with hydrogen sulfide. J. Agric. Food Chem., 1975. 23(3): 501~505
- [35] 孙宝国, 张显卫. 糠硫醇合成的研究. 精细化工, 1993, 10(4): 14~15
- [36] Demole E, Enggist P, Ohloff G. 1-p-Menthene-8-thiol: A powerful flavor impact constituent of grapefruit juice. Helvetica Chimica Acta, 1982, 65(176): 1785~1794
- [37] Swakon E A. Ureas and alcohol products: US, 3317599. 1967.3.2
- [38] Wilhelm P, Sabine W, Tobias V et al. Fragrances and flavor material: US, 6306451. 2001.10.23
- [39] Jones H E. Preparation of β-mercaptoalkanols: US, 3394192. 1968.7.23
- [40] Fletcher J E, Nelson R A. Method of making mercapto alcohols: US, 3462496. 1969.8.19
- [41] Steger B N. Process for making mercapto alcohols: US, 4564710. 1986.1.14
- [42] Shimamoto K, Hatanaka H, Seki S et al. Method for the production of thioalkylene glycols: US, 4493938. 1985.1.15
- [43] Lalancetie J M, Friche A. Reductions with sulfurated borohydrides. VII. Reactions with epoxides. Canndian Journal of Chemistry, 1971, 49: 4047~4053
- [44] Lalancette J M, Laliberte M. 1,2-Dithiols from episulfides. Tetrahedron Letters, 1973, (16): 1401~1404
- [45] Olsen A. Onion-like off-flavor in beer: Isolation and identification of the culprits. Carlsberg Res. Commun., 1988, 53: 1~9
- [46] Robinson P L, Kelly J W, Evans S A. The chemistry of α,ω-mercaptoalcohols in the presence of diethoxytriphenylphosphorane. Temperature dependence of cyclodehydrations and S-ethylations. Phosphorus and Sulfur, 1987, 31: 59~70
- [47] Evers W J, Heinsohn H H, Vock M H. Flavoring with alpha-oxy(oxo)mercaptans: US,

4024289. 1977.5.17

[48] Evers W, Heinsohn H H, Schmitt F L. Alpha-Oxy(oxo)mercaptan perfume and cologne compositions: US, 4070308. 1978.1.24

[49] Brown M D, Whitham G H. α-Thio-substituted ketones as precursors of olefins via oxa-thiolane: Benzyl as protecting group. J. Chem. Soc., Perkin Trans. 1, 1988, (4): 817~821

[50] Ogura K, Furukawa S, Tsuchihashi G I. Mercaptomethylation reaction using cyclohex-anone dimethyl dithioacetal S-oxide. Synthesis Communication, 1976, (3): 202~204

[51] 田红玉, 孙宝国, 徐理阮. 3-巯基-2-丁醇的合成. 化学试剂, 1999, 21(5): 309, 317

[52] Evers W J, Heinsohn H H, Vock M H. Flavoring with α-oxy(oxo)mercaptans: US, 4064278. 1977.12.20

[53] Goldsworthy L J, Harding G F, Norris W L et al. Some sulphides containing the 2-chloroethyl group. J. Chem. Soc., 1948, (4): 2177~2179

[54] Browning G L. Aromatic mercapto-aliphatic ethers as modifiers for butadiene emulsion polymerization: US. 2460567. 1949.2.1

[55] Eliel E L, Doyle T W. Reductions with metal-ammonia combinations. II. Monothioacetals and monothioketals. A synthesis of alkoxy mercaptans. J. Org. Chem., 1970, 35(8): 2716~2722

[56] Eliel E L, Doyle T W, Daignault R A et al. A facile synthesis of β-alkoxy mercaptans and β-thioalkoxy mercaptans. J. Am. Chem. Soc., 1966, 88(8): 1828~1830

[57] Jacobs R L, Schuetz R D. Reaction of 2-alkoxymethylthiiranes with lithium aluminum hydride and with secondary amines. J. Org. Chem., 1961, 26: 3472~3476

[58] Snyder H R, Stewart J M. Production of alkoxy thiols: US, 2497422. 1950.2.14

[59] Snyder H R, Serwart J M, Ziegler J B. The synthesis of mercaptans from olefin sulfides. J. Amer. Chem. Soc., 1947, 69: 2675~2677

[60] Vermeulen C, Collin S. Synthesis and sensorial properties of mercaptoaldehydes. J. Agric. Food Chem., 2002, 50(20): 5654~5659

[61] Vermeulen C, Collin S. Combinatorial synthesis and sensorial properties of 21 mercapto esters. J. Agric. Food Chem., 2003, 51(12): 3618~3622

[62] 章平毅. 新型香料 3-巯基己醇的合成研究//中国香料香精化妆品工业协会. 2004 年中国香料香精学术研讨会论文集. 北京: 中国香料香精化妆品工业协会, 2004. 207~209

[63] Vermeulen C, Christine G D, Collin S. Combinatorial synthesis and sensorial properties of mercapto primary alcohols and analogues. J. Agric. Food Chem., 2003, 51(12): 3623~3628

[64] Vermeulen C, Pellaud J, Gijs L et al. Combinatorial synthesis and sensorial properties of polyfunctional. J. Agric. Food Chem., 2001, 49(11): 5445~5449

[65] 舒宏福. 3-巯基脂肪醇和它的衍生物香料//中国香料香精化妆品工业协会. 2004 年中国香料香精学术研讨会论文集. 北京: 中国香料香精化妆品工业协会, 2004. 190~198

[66] Katz I, Evers W J, Sanderson A. Flavoring compositions and processes utilizing alpha-ketothiols: US, 3773524. 1973.11.20

[67] Greenberg M J. Flavoring with α-mercaptoacetophenone and derivatives: US, 4096284. 1978.6.20

[68] Stacey F W, Harris J F. Radiation-induced addition of hydrogen sulfide to substituted acetylenes. Synthesis of vinylthiol. J. Am. Chem. Soc., 1963, 85: 963~965

[69] May P D, Lee R J. New alkyl dithiol and alkenyl thio: US, 2960538. 1960.11.15

[70] Ford F A. Preparations of mercaptans and sulfides: US, 3045053. 1962.7.17

[71] Lazier W A, Werntz J H. Chemical processes: US, 2402643. 1946.6.25

[72] Signaigo F K. Chemical process: US 2402456. 1946.6.18

[73] Tibler E. Organosulfur compounds from long-chain epoxides. I & EC Product research and development, 1969, 8(4): 415~419

[74] 黄荣初,王兴凤. 肉类香味的合成香料. 有机化学,1983,(3):175~179

[75] Simpson S D. The synthesis of di- and trithiols. Can. J. Research. 1947, 25B: 20~27

[76] Biswell C B. Process for producing mercaptans: US, 2436137. 1948.2.17

[77] Hull C M, Weinland L A, Olsen S R et al. Sulfur bond in vulcanizates. Vulcanization by dithiols. Ind. Eng. Chem., 1948, 40: 513~517

[78] Meade E M, Woodward F N. Some reactions of ethylene sulfide and a new method of preparation of vicinal dithiols. J. Chem. Soc., 1948, 1894~1895

[79] Speziale A J. Ethanedithiol. Org. Syntheses, 1950, 30: 35~37

[80] Grogan C H, Rice L M, Reid E E. Dithiols and derivatives. J. Org. Chem., 1955, 20(1):50~59

[81] Bittell J E, Speier J L. Synthesis of thiols and polysulfides from alkyl halides, hydrogen sulfide, ammonia, and sulfur. J. Org. Chem., 1978,43(9):1687~1689

[82] Societe Nationale des Petroles d'Aquitaine Courbevoie. preparation of alpha-dithiols: DE 2143643. 1972.3.9

[83] Mosciano G et al. Organoleptic characteristics of flavor materials. Perfumer & Flavorist, 1997, 22(3): 47~50

[84] Wilson R A, Giacino C. Flavoring with sulfur containing compounds: US, 3892878. 1975.7.10

[85] Oser B L, Ford R A. Recent progress in the consideration of flavoring ingredients under the food additives amendment, 10. GRAS Substances. Food Technology, 1977, 31(1): 65~74

[86] 孙宝国,田红玉,徐理阮等. 一种 2,3-丁二硫醇的制备方法:中国,ZL00100889.7. 2001.8.22

[87] Hall W P, Reid E E. A series of α,β-dimercaptans. J. Am. Chem. Soc., 1943, 65(2): 1466~1468

[88] 孙宝国,田志敏.1,6-二溴己烷的合成研究.精细化工,1994,11(5):41~42

[89] 孙宝国,李玉,郑福平. 1,6-己二硫醇的合成研究. 精细化工,1996,13(2):17~18

[90] Macleod G. The flavor of beef//Shahidi F. Flavor of Meat and Meat products. New York: Blackie Academic & Professional, 1994. 4~33

[91] Hall R L, Oser B L. Recent progress in the consideration of flavoring ingredients under the food additives amendment. 4. Food Technology, 1970, 24(5): 25~34

[92] Feldman J R, Berg J H. Meat flavor composition and processes: US, 3803330. 1974. 4. 9

[93] Oser B L, Hall R L. Recent progress in the consideration of flavoring ingredients under the food additives amendment, 5. GRAS Substances. Food Technology, 1972, 26(5): 36~41

第四章 硫醚类香料

硫醚类香料是含硫香料中重要的一类,其结构通式为 R—S—R′,式中 R 和 R′可以相同也可以不同。硫醚的化学稳定性比硫醇高,这对于保持香味的质量稳定非常重要。

第一节 硫醚类香料的一般合成方法

一、卤代烃和硫化钾或硫化钠反应

卤代烃和硫化钾或硫化钠反应可以制备对称硫醚。例如[1]:

$$\text{PhCH}_2\text{Cl} + \text{Na}_2\text{S} \longrightarrow \text{PhCH}_2\text{—S—CH}_2\text{Ph}$$

二、硫酸酯和硫化钾或硫化钠反应

硫酸酯和硫化钾或硫化钠反应可以制备对称硫醚。例如:

$$(\text{CH}_3)_2\text{SO}_4 + \text{Na}_2\text{S} \longrightarrow \text{CH}_3\text{—S—CH}_3 + \text{Na}_2\text{SO}_4$$

三、硫醇或硫醇钠与卤代烷反应

硫醇或硫醇钠与卤代烷在碱性条件下反应,可以制备对称硫醚或不对称硫醚。例如[2~4]:

$$\text{(2-甲基-3-巯基呋喃)} + \text{BrCH}_2\text{CH}_2\text{CH(CH}_3\text{)}_2 \xrightarrow{\text{OH}^-} \text{产物}$$

$$\text{PhSH} + \text{BrCH}_2\text{CH}_3 \longrightarrow \text{PhSCH}_2\text{CH}_3$$

$$\text{CH}_3\text{SNa} + \text{ClCH}_2\text{CH}_2\text{OH} \longrightarrow \text{CH}_3\text{SCH}_2\text{CH}_2\text{OH}$$

四、硫醇与烯烃加成

硫醇与烯烃加成是制备不对称硫醚的常用方法。例如[5~7]：

$$\text{furfuryl-CH}_2\text{SH} + \text{(CH}_3)_2\text{C=CH-C(O)CH}_3 \longrightarrow \text{furfuryl-CH}_2\text{-S-C(CH}_3)_2\text{-CH}_2\text{-C(O)CH}_3$$

$$\text{Ph-CH=CH}_2 + \text{PhSH} \longrightarrow \text{Ph-CH=CH-SPh}$$

五、对称二硫醚与烯烃反应

对称二硫醚与烯烃反应可以合成邻二硫醚。例如[8]：

$$\text{PhSSPh} + \text{CH}_2\text{=CH}_2 \xrightarrow[\text{PhMe}]{\text{CpRuCl}} \text{PhS-CH}_2\text{CH}_2\text{-SPh}$$

六、烯丙基溴与对称二硫醚反应

烯丙基溴与对称二硫醚反应可以制备含有烯丙基的不对称硫醚。例如[9]：

$$\text{CH}_2\text{=CH-CH}_2\text{Br} + \text{PhSSPh} \longrightarrow \text{CH}_2\text{=CH-CH}_2\text{-SPh}$$

七、醚和五硫化二磷反应

醚和五硫化二磷反应可以制备硫醚。例如[10]：

$$\text{R-O-R} + \text{P}_2\text{S}_5 \xrightarrow{\triangle} \text{R-S-R} + \text{P}_2\text{O}_5$$

八、含有环氧乙烷环的化合物与硫氰化铵反应

含有环氧乙烷环的化合物与硫氰化铵反应可以转化为相应的含有环硫乙烷环的化合物。例如[11,12]：

$$\text{(cyclohexene oxide)} + \text{NH}_4\text{SCN} \longrightarrow \text{(cyclohexene sulfide)}$$

第二节 简单对称硫醚类香料

简单对称硫醚一般可以用相应的卤代烃和硫化钠经加热制取。例如：

$$\text{CH}_2\text{=CH-CH}_2\text{Cl} + \text{Na}_2\text{S} \longrightarrow \text{(CH}_2\text{=CH-CH}_2)_2\text{S} + \text{NaCl}$$

部分简单对称硫醚类香料的化学结构和香味特征见表 4-1。

表 4-1 部分简单对称硫醚类香料的化学结构和香味特征

名称	化学结构式	香味特征
二甲基硫醚 (FEMA 号 2746)		浆果、洋葱、甘蓝、蔬菜、土豆、西红柿、鱼、扇贝、奶油香气；甜的、奶油、葱、蒜、西红柿、蔬菜味道[13,14]
二丙基硫醚		葱蒜香味[15]
二烯丙基硫醚 (FEMA 号 2042)		洋葱、大蒜、蔬菜、辣根、小萝卜香气和味道[16]
二丁基硫醚 (FEMA 号 2215)	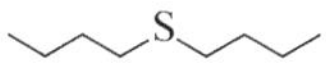	紫罗兰叶青香、花香、洋葱、葱蒜、辣根、蔬菜香气和味道[10]

第三节　简单不对称硫醚类香料

简单不对称硫醚一般可以用硫醇(酚)与卤代烃反应制备。例如：

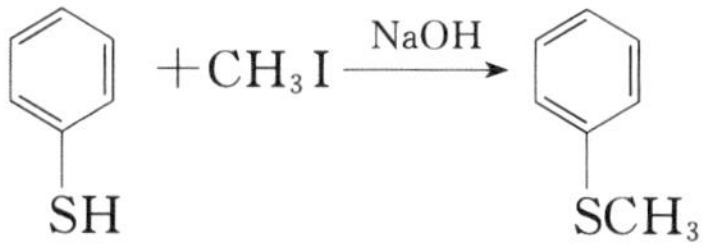

$$\text{PhSH} + \text{CH}_3\text{I} \xrightarrow{\text{NaOH}} \text{PhSCH}_3$$

部分简单不对称硫醚类香料的化学结构和香味特征见表 4-2。

表 4-2 部分简单不对称硫醚类香料的化学结构和香味特征

名称	化学结构式	香味特征
甲基乙基硫醚 (FEMA 号 3860)		蒜、西红柿、甘蓝、蔬菜、肉香、咖啡、吡啶香气；蔬菜、谷物、西红柿、蛋、肉香、蛤肉、血味、洋葱、大蒜味道[10]
甲基苯基硫醚 (FEMA 号 3873)	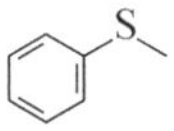	辛香、木香香气；木香、烤咖啡味道[10]

第四节　糠基硫醚类香料

一、糠基甲基硫醚

糠基甲基硫醚(furfuryl methyl sulfide，FEMA 号 3160)为浅黄色液体，天然存在于香油、咖啡中，具有洋葱、大蒜、辣根香气和味道，用于调味品、咖啡、洋葱、大蒜、辣根、芥末等食品香精[17,18]。

糠基甲基硫醚可以用糠硫醇与硫酸二甲酯反应以几乎定量完成的产率制备[19]。

$$\text{furyl-CH}_2\text{SH} + (\text{CH}_3)_2\text{SO}_4 \longrightarrow \text{furyl-CH}_2\text{SCH}_3$$

二、糠基异丙基硫醚

糠基异丙基硫醚(furfuryl isopropyl sulfide, FEMA 号 3161)为无色至浅黄色液体,长时间放置颜色加深,具有葱、蒜、咖啡香气和味道,可用于洋葱、大蒜、咖啡及肉味等食品香精。

糠基异丙基硫醚可以用糠硫醇与 2-溴丙烷反应制备[10]。

$$\text{furyl-CH}_2\text{SH} + \text{Br-CH(CH}_3)_2 \xrightarrow{\text{NaOH}} \text{furyl-CH}_2\text{-S-CH(CH}_3)_2$$

三、二糠基硫醚

二糠基硫醚(difurfuryl sulfide, FEMA 号 3238)为无色液体,天然存在于咖啡、香油中。具有咖啡、肉香香气,蘑菇味道,可用于咖啡、可可、芝麻油、蘑菇等食品香精。

二糠基硫醚可以用糠硫醇与糠基溴或糠基氯反应制备[20]。

$$\text{furyl-CH}_2\text{SH} + \text{furyl-CH}_2\text{Br} \xrightarrow{\text{NaOH}} \text{furyl-CH}_2\text{SCH}_2\text{-furyl}$$

糠基溴可以通过糠醇与三溴化磷反应制备[21];糠基氯可以通过糠醇与氯化亚砜反应制备[22]。

四、其他糠基硫醚香料

部分其他糠基硫醚类香料的化学结构和香味特征见表 4-3。

表 4-3 部分其他糠基硫醚类香料的化学结构和香味特征

名称	化学结构式	香味特征
糠基乙基硫醚	furyl-CH$_2$-S-C$_2$H$_5$	肥肉香、青香、熟蒜香、蘑菇香[23]
糠基丙基硫醚	furyl-CH$_2$-S-C$_3$H$_7$	烤面包香、蘑菇香[23]

续表

名称	化学结构式	香味特征
糠基丁基硫醚		蘑菇香、青香[23]
糠基异丁基硫醚		炒洋葱香、煮洋葱香、小茴香香、香肠香、青香[23]
糠基戊基硫醚		果香、梨香[23]
糠基异戊基硫醚		大蒜香、韭葱香、芸香、果香[23]
糠基庚基硫醚		药草香、咖啡、菊苣香[23]
糠基苄基硫醚		焦咖啡香、橡胶香味[23]
糠基丙烯基硫醚		鱼香、白葡萄酒香、鲱鱼香、大蒜香、花香[23]
糠基 2-氯丙基硫醚		大蒜香、蘑菇香、药草香、细香葱香[23]
1,2-二糠硫基乙烷		咖啡香、油炸吐司面包香、蘑菇蔬菜香[23]
1,3-二糠硫基醚丙烷		蔬菜香、香叶香、焙烤香、大蒜香、伏特加酒香、鱼香[23]

第五节 3-呋喃硫醚类香料

一、2-甲基-3-甲硫基呋喃

2-甲基-3-甲硫基呋喃是于 2000 年才获得 FEMA 号的新香料,FEMA 号 3949,在肉制品中的推荐用量为 $0.05\sim0.5$ mg/kg,在焙烤食品中的推荐用量为 $0.02\sim0.2$ mg/kg[24]。1986 年,Macleod 和 Ames 在炖牛肉挥发性香成分中发现了该化合物,随后在酵母提出物和咖啡香成分也发现了该化合物[25]。该香料在 $<$ 1μg/kg 时具有肉香;在 $>$ 1μg/kg 时具有维生素 B_1 样香味。其香气阈值为 50ng/kg,香味阈值为 5ng/kg[26-28]。

2-甲基-3-甲硫基呋喃可以用 2,3-二溴-2-烯-1,4-丁二醇为起始原料合成[29]。

[反应式: HOCH₂C(Br)=C(Br)CH₂OH $\xrightarrow{H_2SO_4}$ 3,4-二溴呋喃 $\xrightarrow{Li^tBu}$ 3-溴-4-锂呋喃 $\xrightarrow{CH_3SSCH_3}$ 3-溴-4-甲硫基呋喃 $\xrightarrow{Li^tBu}$ 3-锂-4-甲硫基呋喃 $\xrightarrow{H_2O}$ 3-甲硫基呋喃 $\xrightarrow{(CH_3)_2SO_4}$ 2-甲基-3-甲硫基呋喃]

2-甲基-3-甲硫基呋喃也可以用 2-甲基-3-巯基呋喃和碘甲烷[2]或硫酸二甲酯合成[30]。

[反应式: 2-甲基-3-巯基呋喃 + CH₃I $\xrightarrow{OH^-}$ 2-甲基-3-甲硫基呋喃]

2-甲基-3-甲硫基呋喃还可以用 4-羰基-2-戊烯醛为原料合成[31]。

[反应式: CH₃COCH=CHCHO $\xrightarrow{CH_3SH}$ CH₃COCH(SCH₃)CH₂CHO → 2-甲基-3-甲硫基呋喃]

二、其他一些 3-呋喃硫醚类化合物

其他一些 3-呋喃硫醚类香料的化学结构和香味特征见表 4-4。

表 4-4 其他一些 3-呋喃硫醚类香料的化学结构和香味特征

名称	化学结构式	香味特征
2-甲基-3-乙硫基呋喃	[结构式]	肉香[32]
2-甲基-3-(1-巯基-1-乙硫基)呋喃	[结构式]	肉香[32]
2,5-二甲基-3-丙硫基呋喃	[结构式]	烤肉香[33]
2,5-二甲基-3-(1-甲基-2-巯基丙硫基)呋喃	[结构式]	牛肉清汤味、咸猪肉味[34]
2-甲基-3-烯丙硫基呋喃	[结构式]	洋葱香、大蒜香气;鱼香、金属、细香葱香味;青香、洋葱、焦香香韵[35]

第六节 2-烷硫基吡啶类香料

一、2-烷硫基吡啶类化合物的一般合成方法

2-烷硫基吡啶类化合物可以通过 2-吡啶硫酮在相转移催化下与卤代烷反应制备[23]。

二、2-烷硫基吡啶类化合物

一些 2-烷硫基吡啶类香料的化学结构和香味特征见表 4-5。

表 4-5 一些 2-烷硫基吡啶类香料的化学结构和香味特征[23]

名称	化学结构式	香味特征
2-甲硫基吡啶		辛香、醚香、壤香、吡啶样
2-乙硫基吡啶		椰子香、生橡胶样、霉味、鲜广藿香、鲜紫罗兰香
2-丙硫基吡啶		木香、甘草香、泥土清香、果香
2-异丙硫基吡啶		木香、泥土清香
2-丁硫基吡啶		泥土清香、果香
2-异丁硫基吡啶		塑料气息、乳香
2-异戊硫基吡啶		大茴香样、泥土清香、金属样、熟蔬菜香
2-戊硫基吡啶		小茴香样、甜的青香
2-庚硫基吡啶		臭鸡蛋、茉莉酮样、壤香、蘑菇香

续表

名称	化学结构式	香味特征
1,2-二(2-吡啶硫基)乙烷		甘草香、淡羔羊肉香、泥土清香、木香、青叶香
1,3-二(2-吡啶硫基)丙烷		甘草香、大茴香香、茉莉酮样

第七节 2-烷硫基嘧啶类香料

一、2-烷硫基嘧啶类化合物的一般合成方法

2-烷硫基嘧啶类化合物可以通过 2-巯基嘧啶在相转移催化下与卤代烷反应制备[23]。

$$\text{嘧啶-SH} + RX \longrightarrow \text{嘧啶-SR} + HX$$

二、2-烷硫基嘧啶类化合物

一些 2-烷硫基嘧啶类香料的化学结构和香味特征见表 4-6。

表 4-6 一些 2-烷硫基嘧啶类香料的化学结构和香味特征[23]

名称	化学结构式	香味特征
2-甲硫基嘧啶		甘蓝香、烤香
2-乙硫基嘧啶		甘蓝香、旱芹块根香
2-丙硫基嘧啶		野蒜香、阿拉伯李子香、旱芹香
2-异丙硫基嘧啶		熟蒜香、旱芹块根香、薄荷香
2-丁硫基嘧啶		大蒜香、薄荷香、大茴香香

续表

名称	化学结构式	香味特征
2-异丁硫基嘧啶	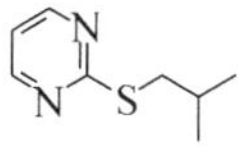	薄荷香
2-戊硫基嘧啶	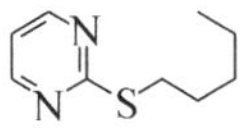	菊苣香、烂铁气味、苦味
2-异戊硫基嘧啶	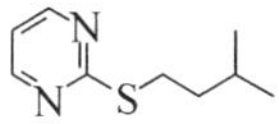	酒花香、泥土清香、啤酒香
2-庚硫基嘧啶	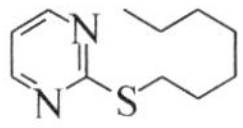	烂铁气味
2-烯丙硫基嘧啶	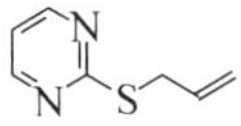	辛辣、洋葱香、鲜蒜香

第八节 β-羟基硫醚类香料

一、β-羟基硫醚香料化合物的一般合成方法

1. α,β-环氧化物与含硫试剂的开环加成反应

各种含硫试剂在催化剂的作用下与环氧乙烷或其衍生物进行开环加成反应合成 β-羟基硫醚类化合物是比较实用的方法。含硫试剂有硫醇、含硫硅烷试剂等。

(1) 硫醇作为亲核试剂

硫醇与环氧乙烷可在催化剂作用下进行开环反应,催化剂主要有碱[36,37]、季铵盐[38,39]、蒙脱石黏土[40]、镓锂络合物[41]、硝酸铈铵[42]、聚乙二醇[43]、三价铬和四价钛络合物[44]、活性氧化铝[45,46]、沸石[47]以及各种路易斯酸,如 InCl$_3$[48]、LnCl$_3$(Ln 为 Ce,Sm)[49]、CoCl$_2$ 或(Co$_2$)CO[50]、SmI$_2$[51]、酒石酸二价金属盐[52,53]等。

此反应操作简单,产率高,是合成 β-羟基硫醚比较实用的方法之一[54~56]。

(2) 其他的含硫亲核试剂

含硫硅烷试剂如 PhSSiMe$_3$[57,58]、Ph$_3$SiSH[59,60],在相应催化剂的作用下通过亲核开环和水解反应也可制取 β-羟基硫醚。

$$\text{环氧化合物} + R_1SSiR_2 \xrightarrow{\text{催化剂}} \xrightarrow{\text{水解}} \underset{OH \quad SR_1}{\text{产物}}$$

其他的含硫亲核试剂还有 $RSBH_2^{[61,62]}$、$PhSCH(OMe)Li^{[63]}$、$RSLi^{[64,65]}$ 等。

2. 羰基化合物与含硫试剂的亲核加成反应

含硫亲核试剂与羰基加成常用的含硫试剂有 $ArSCH_2SiMe_3^{[66,67]}$、$RSCH=CHLi^{[68,69]}$、$[PhSCR_1R_2]Li^{[70\sim72]}$、$RSCH_2Li^{[73,74]}$ 等。例如：

$$\underset{R_2}{\overset{R_1}{>}}\!\!=\!\!O + \text{含硫试剂} \xrightarrow{\text{加成}} \xrightarrow{\text{水解}} \underset{OH \quad SR}{\overset{R_1 \quad R_3}{\underset{R_2 \quad R_4}{}}}$$

R_3、R_4 为 H 或烷基

3. α-烷硫基羰基化合物的还原反应

α-烷硫基羰基化合物经过还原剂还原可制取 β-羟基硫醚[75~91]。常用的还原剂有 $NaBH_4^{[75\sim77]}$、$LiAlH_4^{[77\sim91]}$、$Zn(BH_4)_2^{[82]}$、$BuLi^{[83]}$、$RMgBr^{[83]}$、$Et_2Zn^{[84]}$、酵母[86,87]等。

$$R_1\underset{SR_2}{\overset{O}{\underset{|}{C}}}\!\!-\!\!\underset{}{\overset{R_3 \, R_4}{C}} \xrightarrow{[H]} R_1\underset{SR_2}{\overset{OH}{\underset{|}{C}}}\!\!-\!\!\underset{}{\overset{R_3 \, R_4}{C}}$$

4. 1,3-氧硫杂环戊烷的开环反应

1,3-氧硫杂环戊烷在还原剂作用下，开环生成 β-烷硫基乙醇[92~96]。常用的还原剂有 $SmI_2^{[92]}$、$LiAlH_4^{[95]}$ 等。

$$\underset{S}{\overset{O}{\diagup\!\!\!\diagdown}}\!\!\underset{R_2}{\overset{R_1}{<}} \xrightarrow{[H]} R_1\underset{}{\overset{R_2}{\underset{|}{C}}}\!\!-\!\!S\!\!-\!\!CH_2CH_2OH$$

5. β-巯基醇硫醚化反应

在碱存在下，卤代烃与巯基醇发生硫醚化反应也可以生成 β-羟基硫醚[97~100]。

$$RX + \underset{SH \, OH}{\bowtie} \xrightarrow{OH^-} \underset{SR \, OH}{\bowtie}$$

巯基醇和硫酸二甲酯[101]、甲基烯丙基碳酸酯[102]、Me_3SeOH[103]发生硫醚化反应,也可以生成 β-羟基硫醚。

巯基醇与烯烃[104]发生加成反应也生成 β-羟基硫醚。

6. β-卤代醇与硫醇的亲核取代反应

β-卤代醇和硫醇在碱的存在下发生亲核取代反应可生成 β-羟基硫醚[105～110]。

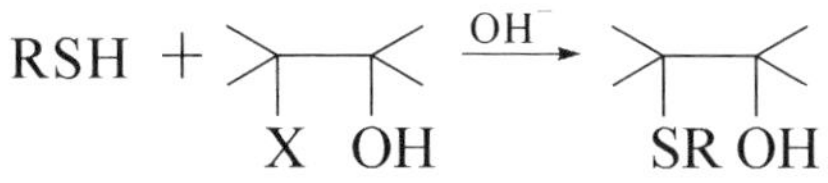

$$RSH + \underset{X\ OH}{\diagdown\diagup} \xrightarrow{OH^-} \underset{SR\ OH}{\diagdown\diagup}$$

二、β-羟基硫醚香料化合物

一些 β-羟基硫醚香料的化学结构和香味特征见表 4-7。

表 4-7　一些 β-羟基硫醚香料的化学结构和香味特征

名称	化学结构式	香味特征或应用建议
2-甲硫基乙醇（FEMA 号 4004）	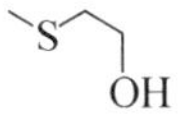	可用于焙烤食品、软饮料、糖果、肉制品、奶制品、调味品等香精[111]
1-乙硫基-2-丁醇	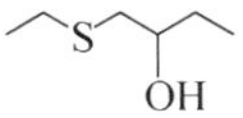	大蒜香气,带有葱香和蔬菜香气
1-丙硫基-2-丁醇	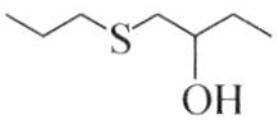	葱蒜香气,带青的果肉香气
1-丁硫基-2-丁醇	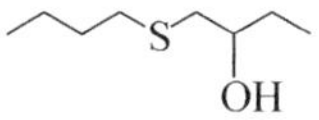	葱香,带有蒜香
1-己硫基-2-丁醇	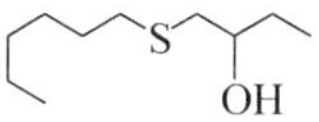	青涩的果香,带葱蒜香韵
1-辛硫基-2-丁醇	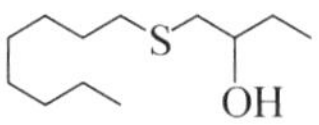	橡胶样香气,带有香芹酮气息
1-糠硫基-2-丁醇	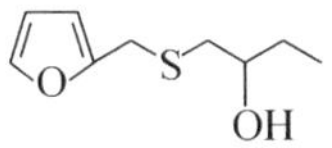	热带水果香气,略带葱蒜香气
1-烯丙硫基-2-丁醇	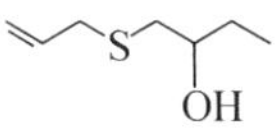	腌制过的萝卜香气,带葱蒜香气
3-丙硫基-4-庚醇	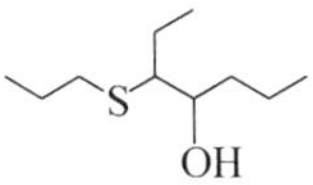	紫罗兰叶、甜瓜、黄瓜、青香、果香、蔬菜、花香香气;紫罗兰叶、甜瓜、黄瓜、青香、柑橘、蔬菜、大蒜香味[75]

第九节 β-烷氧基硫醚类香料

β-烷氧基硫醚可以由相应的 β-烷硫基醇通过醚化制备。例如[112]：

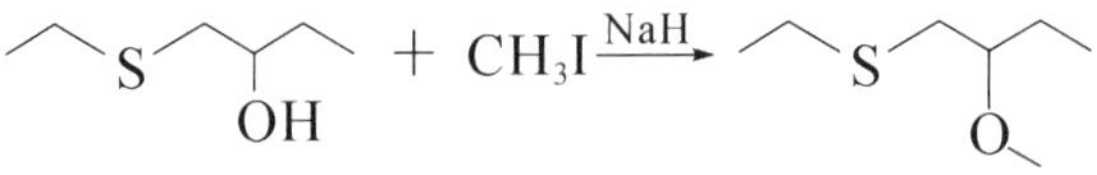

$$\text{CH}_3\text{CH}_2\text{S}\underset{\text{OH}}{\diagdown}\text{CH}_2\text{CH}_3 + \text{CH}_3\text{I} \xrightarrow{\text{NaH}} \text{CH}_3\text{CH}_2\text{S}\underset{\text{O}}{\diagdown}\text{CH}_2\text{CH}_3$$

一些 β-烷氧基硫醚类香料的化学结构和香味特征见表 4-8。

表 4-8 一些 β-烷氧基硫醚类香料的化学结构和香味特征

名称	化学结构式	香味特征
2-甲氧基丁基乙硫醚	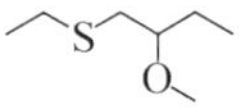	青、涩的蔬菜香,略带葱蒜香气
2-甲氧基丁基丙硫醚	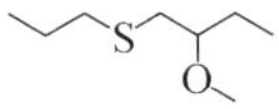	甜、涩的萝卜香气
2-甲氧基丁基丁硫醚	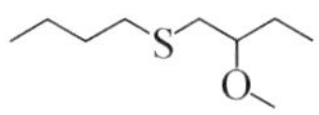	甜的萝卜香气,带有葱香香气
2-甲氧基丁基己硫醚	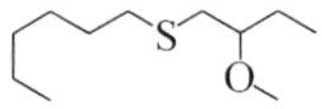	金属样的香气,略带葱蒜香气
2-甲氧基丁基辛硫醚	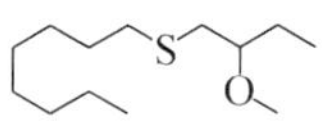	煮过的蔬菜香气
2-甲氧基丁基糠硫醚	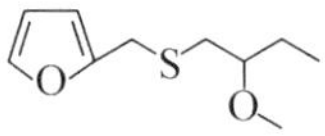	鲜咸香气,带葱蒜香气

第十节 β-巯基硫醚类香料

一、β-巯基硫醚香料化合物的一般合成方法

1. β-烷硫基卤代烃的取代反应

β-烷硫基卤代烃中的卤原子被 SH 基团取代,即可生成相应的硫醇。如用 2-烷硫基卤代乙烷与硫氢化钾溶液反应生成 2-烷硫基乙硫醇[113]。

$$\text{RS}\diagdown\text{X} + \text{KSH} \longrightarrow \text{RS}\diagdown\text{SH}$$

式中:R 为甲基、乙基、丙基、丁基。

2. 1,3-二硫杂环戊烷的开环反应

(1) 在金属和液氨存在下反应

1,3-二硫杂环戊烷在金属和液氨存在下,发生开环反应生成 β-巯基硫醚[114~116]。

$$\begin{array}{c}\text{S}\\ \text{S}\end{array}\!\!\!\!\bigg\langle\!\!\!\begin{array}{c}R_1\\ R_2\end{array} \xrightarrow[\text{液氨}]{M} R_1\!\!-\!\!\underset{R_2}{\overset{}{\text{C}}}\!\!-\!\!S\!\!-\!\!CH_2CH_2SH$$

式中:M 为 K、Na、Li、Ca,其中 Ca 的催化效果最好。

(2) 在吡啶-BH_3-$AlCl_3$ 作用下反应

1,3-二硫杂环戊烷在吡啶-BH_3-$AlCl_3$ 作用下,在溶剂 CH_2Cl_2 中反应可开环生成相应的 β-巯基硫醚[117]。

$$\begin{array}{c}\text{S}\\ \text{S}\end{array}\!\!\!\!\bigg\langle\!\!\!\begin{array}{c}R_1\\ R_2\end{array} \xrightarrow[CH_2Cl_2]{\text{吡啶-}BH_3\text{-}AlCl_3} R_1\!\!-\!\!\underset{R_2}{\overset{}{\text{C}}}\!\!-\!\!S\!\!-\!\!CH_2CH_2SH$$

3. β-羟基硫醚与硫脲的反应

烃硫基醇与硫脲、盐酸作用后用碱水解,可生成相应的 β-巯基硫醚[118]。

$$RS\!\!-\!\!CH_2CH_2OH + HCl + \underset{H_2N}{\overset{S}{\underset{\|}{C}}}\!\!NH_2 \longrightarrow$$

$$RS\!\!-\!\!CH_2CH_2\!\!-\!\!S\!\!-\!\!\underset{NH_2}{\overset{NH}{\underset{\|}{C}}}\cdot HCl \xrightarrow{NaOH} RS\!\!-\!\!CH_2CH_2SH$$

4. 硫杂环丙烷的开环反应

硫杂环丙烷在催化剂的作用下,与硫醇反应,开环生成 β-巯基硫醚。催化剂可以是三氟化硼、BF_3 与乙醚或乙酸的复合物、沸石、醇钠、氢氧化钠、三乙胺、硫酸等[119~123]。

$$\underset{S}{\bigtriangleup} + RSH \xrightarrow{\text{催化剂}} \underset{SH\;SR}{\diagdown\!\!\diagup}$$

5. β-巯基胺化合物与硫醇铵反应

β-巯基胺化合物与硫醇铵（NH_4SR）加热反应可制得 β-巯基硫醚化合物[124,125]。反应所用溶剂为极性溶剂，如水、醇、氨、胺、二甲基砜、N,N-二甲基甲酰胺、四氢呋喃、乙腈等。

$$\begin{array}{c} RR \\ R-C-C-R \\ SHN(R)_2 \end{array} + NH_3 + R'SH \longrightarrow \begin{array}{c} RR \\ R-C-C-R \\ SHSR' \end{array}$$

式中：R 和 R' 为 $C_1 \sim C_8$ 烃基。

二、β-巯基硫醚香料化合物的香味特征

一些 β-巯基硫醚类香料的化学结构和香味特征见表 4-9。

表 4-9　一些 β-巯基硫醚类香料的化学结构和香味特征

名称	化学结构式	香味特征
1-乙硫基-2-丁硫醇		浓郁的葱蒜香，有蔬菜香韵
1-丙硫基-2-丁硫醇		似留兰香香气
1-丁硫基-2-丁硫醇		似留兰香香气，带瓜香
1-己硫基-2-丁硫醇		烤香，带瓜香
1-辛硫基-2-丁硫醇		焙烤过的谷物香
1-糠硫基-2-丁硫醇		炒米香

第十一节　β-羰基硫醚类香料

一、1-甲硫基-2-丙酮

1-甲硫基-2-丙酮（1-methylthio-2-propanone，FEMA 号 3882）为无色至浅黄色液体，具有金枪鱼罐头香气和味道，可用于咖啡、奶制品、焙烤食品、软饮料、调味品、肉味等食品香精[126]。

1-甲硫基-2-丙酮可以由氯丙酮与甲硫醇钠反应制备[10]。

$$\text{CH}_3\text{COCH}_2\text{Cl} + \text{CH}_3\text{SNa} \longrightarrow \text{CH}_3\text{COCH}_2\text{SCH}_3 + \text{NaCl}$$

用烷硫醇钠代替甲硫醇钠,可以合成其他 1-烷硫基-2-丙酮化合物[127,128]。

二、1-甲硫基-2-丁酮

1-甲硫基-2-丁酮(1-methylthio-2-butanone,FEMA 号 3207)是咖啡香成分,为无色液体,具有近似于蘑菇的气味,并带有一种特殊的大蒜底香,常用于洋葱、大蒜、蘑菇等食品香精。

1-甲硫基-2-丁酮可以由 1-氯-2-丁酮与甲硫醇钠反应制备[10]。

$$\text{CH}_3\text{CH}_2\text{COCH}_2\text{Cl} + \text{CH}_3\text{SNa} \longrightarrow \text{CH}_3\text{CH}_2\text{COCH}_2\text{SCH}_3 + \text{NaCl}$$

一些 β-羰基硫醚类香料的化学结构和香味特征见表 4-10。

表 4-10 一些 β-羰基硫醚类香料的化学结构和香味特征

名称	化学结构式	香味特征
1-乙硫基-2-丁酮		浓郁的青萝卜香气、带葱蒜香气
1-丙硫基-2-丁酮		浓郁的萝卜干香气
1-丁硫基-2-丁酮		葱蒜香气、带炖肉香气
1-己硫基-2-丁酮		橡胶样香气、有腌制过的蔬菜香气、带炖肉香气
1-辛硫基-2-丁酮		腌制过的蔬菜香气、有蘑菇香韵
1-糠硫基-2-丁酮		烤坚果香气、似焙炒过的大麦香气
1-烯丙硫基-2-丁酮		浓郁的青香、蔬菜香气
3-甲硫基-4-庚酮		青香、松木香、果香、黑醋栗香气;甜的、果香、黑醋栗香、薄荷香味[75]

名称	化学结构式	香味特征
3-丙硫基-4-庚酮		甜的、花香、茉莉香、浆果样香气;甜的、花香、茉莉香、葡萄柚香、黑醋栗香味[75]
3-异丁硫基-4-庚酮		甜的、花香、柑橘香、果香香气;青香、花香、薄荷香、果香、柑橘香味[75]
3-甲基烯丙硫基-2,6-二甲基-4-庚酮		柑橘香、葡萄柚香、花香、玫瑰香气;柑橘香、葡萄柚香、花香、辛香、青香、果香香味[75]
3-(2-烯丁硫基)-2,6-二甲基-4-庚酮		花香、柑橘香、葡萄柚香、木香香气;柑橘、葡萄柚香味[75]
3-(4-酮-3-庚硫基)-2,6-二甲基-4-庚酮		葡萄柚香、花香、木香香气;甜的、硫化物样、葡萄柚香味[75]

第十二节 γ-羟基硫醚类香料

γ-羟基硫醚香料有3-甲硫基丙醇。

3-甲硫基丙醇(3-methylthiopropanol,FEMA号3415)俗称菠萝醇,天然存在于啤酒、威士忌、白葡萄酒中。为浅黄色透明液体,具有甜的、洋葱、大蒜、菜肴、汤、蔬菜香气以及洋葱、大蒜、菜肴、牛肉清汤味道,常用于调配西红柿、酱油、奶酪、蔬菜、肉汤、肉味等食品香精。

3-甲硫基丙醇可以由3-氯丙醇与甲硫醇钠反应制备[129]。

$$Cl\text{—}CH_2CH_2CH_2\text{—}OH + CH_3SNa \longrightarrow CH_3S\text{—}CH_2CH_2CH_2\text{—}OH$$

也可以由3-甲硫基丙醛还原制备。

$$CH_3S\text{—}CH_2CH_2\text{—}CHO + NaBH_4 \longrightarrow CH_3S\text{—}CH_2CH_2CH_2\text{—}OH$$

第十三节 γ-羰基硫醚类香料

一、3-甲硫基丙醛

3-甲硫基丙醛(3-methylthiopropionaldehyde,FEMA 号 2747)俗称菠萝醛,天然存在于西红柿、威士忌、乳酪、土豆中。为无色至浅黄色液体,具有大蒜、洋葱、土豆、西红柿、壤香、蔬菜、肉汤香气以及脂肪、土豆、西红柿、蔬菜、牛肉汤、鸡蛋、海鲜味道,常用于调配土豆、西红柿、肉、海鲜、洋葱、面包、坚果、奶酪、热带水果、蔬菜等食品香精。

3-甲硫基丙醛可以由丙烯醛与甲硫醇反应制备。

$$CH_3SH + \diagup\!\!\!\diagdown CHO \longrightarrow \diagup\!\!\!S\!\!\!\diagdown CHO$$

二、3-甲硫基丁醛

3-甲硫基丁醛(3-methylthiobutanal,FEMA 号 3374)为无色至浅黄色液体,具有西红柿、甘蓝、蔬菜、青香、鱼香、奶酪香气以及蔬菜、西红柿、鱼香味道,用于调配甘蓝、西红柿、土豆、鱼类等食品香精。

3-甲硫基丁醛常由巴豆醛与甲硫醇反应制备[130]。

$$\diagup\!\!\!\diagdown CHO + CH_3SH \longrightarrow \diagup\!\!\!\diagdown_S\!\!\!\diagdown CHO$$

三、4-甲硫基-2-丁酮

4-甲硫基-2-丁酮(4-methylthio-2-butanone,FEMA 号 3375)为无色液体,具有西红柿、奶酪、鱼香、蔬菜、土豆、脂肪香气以及油腻、鱼香、土豆、蔬菜、蘑菇味道,常用于调配土豆、西红柿、菠萝、鱼、蘑菇香精。沸点 106℃/55mmHg①、77~78℃/20mmHg,折光率 n_D^{20} 1.4711。

4-甲硫基-2-丁酮可以由甲基乙烯基酮与甲硫醇反应制备[130]。

四、4-甲硫基-4-甲基-2-戊酮

4-甲硫基-4-甲基-2-戊酮（4-methylthio-4-methyl-2-pentanone,FEMA 号

---

① 1mmHg=1.333 22×10² Pa,下同。

3376)为无色液体,具有青香、霉香、热带水果、油炸大蒜香气以及青香、辣根、蔬菜、药草、油炸大蒜味道,常用于甜瓜、菠萝、桃子、芥菜、蔬菜、咖啡、浆果、热带水果、调味品、肉味等食品香精。沸点78℃/15mmHg,折光率n_D^{13}1.4750。

4-甲硫基-4-甲基-2-戊酮可以由异丙叉丙酮与甲硫醇反应制备[10]。

其他一些γ-羰基硫醚类香料的化学结构和香味特征见表4-11。

表4-11 其他一些γ-羰基硫醚类香料的化学结构和香味特征

名称	化学结构式	香味特征
2-甲硫基甲基-2-丁烯醛		脂肪香,土豆香[32]
1-甲硫基-3-戊酮		甜的,牛奶香,奶油香,黄油香,烤香[32]
4-甲基-4-乙硫基-2-戊酮		壤香,似辣根香气
4-甲基-4-丙硫基-2-戊酮		青、涩的萝卜香气
4-甲基-4-丁硫基-2-戊酮		甜、涩的萝卜香气,带有木香香气
对 -5-甲硫基-3-酮		大蒜香,干酪香,薄荷香,黑加仑子香味[23]
对 -5-丁硫基-3-酮		大蒜香,青香,黑加仑子香,药草香,辛香[23]
对 -5-戊硫基-3-酮		药草香,果香,薄荷香,黑加仑子香,青香[23]

续表

名称	化学结构式	香味特征
对-5-异戊硫基-3-酮	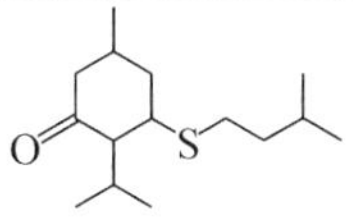	干酪香,万寿菊香,大茴香香,黑加仑子香,青香,薄荷香,焦牛肉香味[23]
对-5-庚硫基-3-酮	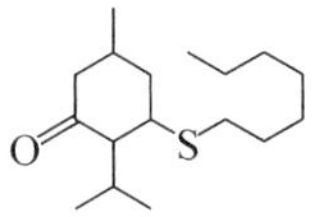	蘑菇香,青香,薄荷香味[23]

第十四节 β,γ-二羰基硫醚类香料

β,γ-二羰基硫醚可以通过下面的方法合成[131]:

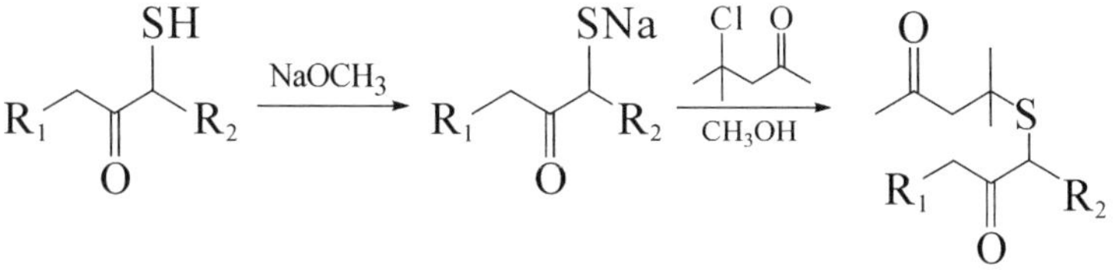

一些 β,γ-二羰基硫醚化合物的化学结构和香味特征如下:

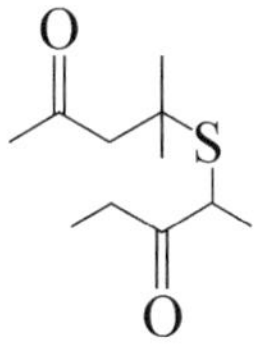

醚香,稍带有微弱木香香气[131]

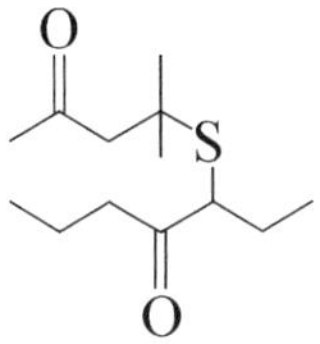

醚香为主,伴有木香和酸香香气[131]

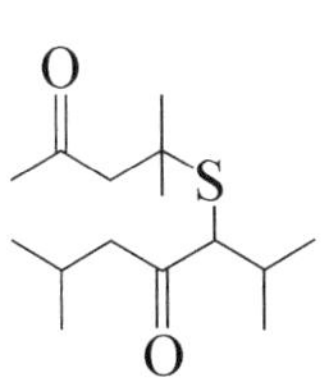

以酸香贯穿整个香气[131]

第十五节 3-烷硫基酯类香料

最简单的 3-烷硫基酯是 3-甲硫基丙酸甲酯(FEMA 号 2720),天然存在于白葡萄酒、菠萝等香成分中,具有蔬菜、西红柿、辣根、洋葱、大蒜、肉香气以及蔬菜、小萝卜、辣根、菠萝味道,可用于猪肉、牛肉、鸡肉、辣根、洋葱、大蒜、西红柿、菠萝等食品香精[132]。

3-甲硫基丙酸甲酯可以通过下面的方法制备:

(1) 甲硫醇钠和 3-溴丙酸甲酯反应[10]

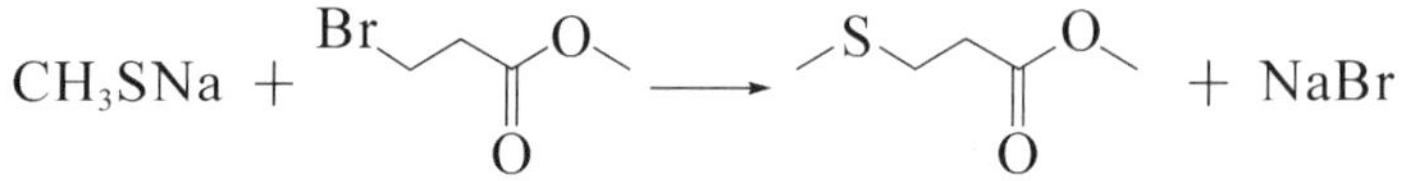

$$CH_3SNa + \text{Br-CH}_2CH_2COOCH_3 \longrightarrow CH_3S\text{-CH}_2CH_2COOCH_3 + NaBr$$

(2) 甲硫醇和丙烯酸甲酯反应[133]

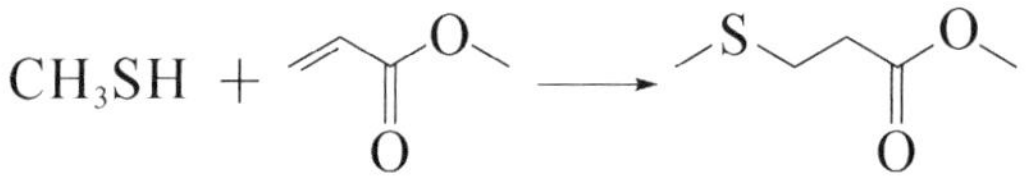

$$CH_3SH + CH_2=CHCOOCH_3 \longrightarrow CH_3S\text{-CH}_2CH_2COOCH_3$$

其他一些 3-烷硫基酯类化合物都可以用类似的方法制备[134]。

其他一些 3-烷硫基酯类香料的化学结构和香味特征见表 4-12。

表 4-12 一些 3-烷硫基酯类香料的化学结构和香味特征

名称	化学结构式	香味特征与应用建议
3-甲硫基丙酸乙酯 (FEMA 号 3343)	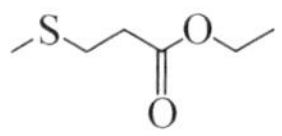 $CH_3SCH_2CH_2COOC_2H_5$	水果、煮菠萝、西红柿香气;菠萝、洋葱、大蒜、蔬菜、朗姆酒味道
3-甲硫基丁酸乙酯 (FEMA 号 3836)	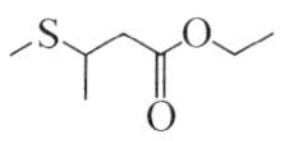 $CH_3SCH(CH_3)CH_2COOC_2H_5$	可用于焙烤食品、肉制品、调味品等香精[126]
3-糠硫基丙酸乙酯 (FEMA 号 3674)	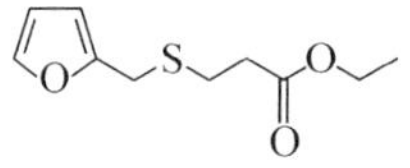 糠基-S-CH$_2$CH$_2$COOC$_2$H$_5$	洋葱香、大蒜香、咖啡香、蔬菜香气;10mg/kg 烤香、坚果香、咖啡香味道[17,18]。可用于焙烤食品、饮料、糖果、奶制品等香精[135]

第十六节 1,4-二噻烷类香料

一、1,4-二噻烷

1,4-二噻烷(1,4-dithiane,FEMA 号 3831)为白色结晶,具有海鲜、大蒜、洋葱、吡啶样香气,洋葱、蒜、蘑菇、海鲜、蛤肉、牡蛎、火鸡肉味道,可用于调配辣根、蘑菇、寿司、贝类海鲜、蛤肉、牡蛎、火鸡等食品香精[126]。

1,4-二噻烷可以用1,2-乙二硫醇和1,2-二溴乙烷在乙醇钠的存在下反应制得[136,137]。

$$\begin{Bmatrix}SH\\SH\end{Bmatrix} + \begin{Bmatrix}Br\\Br\end{Bmatrix} \xrightarrow{C_2H_5ONa} \begin{matrix}S\\S\end{matrix} + 2HBr$$

$HOCH_2CH_2SH$ 在低温下,用 Bz_2O_2 催化并在紫外线照射下与 $CH_2=CHCl$ 反应制取 $HOCH_2CH_2SCH_2CH_2Cl$ 时,在蒸馏产物的过程中,残留物中有大量的 1,4-二噻烷[138]。

将 $HSCH_2CH_2OH$ 加热到 160~185℃ 反应 3h,1,4-二噻烷的收率就能达到 39%[139]。

$$2HSCH_2CH_2OH \xrightarrow[3h]{160\sim185℃} \begin{matrix}S\\S\end{matrix} + 2H_2O$$

$(HOCH_2CH_2)_2S$ 在 N_2 存在和 Al_2O_3 催化下,在 225℃ 时反应 90min 可以生成 1,4-二噻烷[140]。

环氧乙烷、环硫乙烷等在 Al_2O_3 催化下,可以生成 1,4-二噻烷[141~144]。

$$\overset{O}{\triangle} + H_2S \xrightarrow[200℃]{Al_2O_3} \begin{matrix}S\\S\end{matrix}$$

$$\overset{S}{\triangle} \xrightarrow[N_2]{Al_2O_3} H_2S + C_2H_4 \xrightarrow{220℃} \begin{matrix}S\\S\end{matrix}$$

$$(CH_3)_3CS(CH_2)_2Cl \xrightarrow[235\sim345℃]{Al_2O_3,N_2} \begin{matrix}S\\S\end{matrix}$$

在其他催化剂作用下合成1,4-二噻烷的方法有[145,146]:

$$C_2H_4 + H_2S + CS_2 \xrightarrow[MoO_2]{CoO} \begin{matrix}S\\S\end{matrix}$$

$$\overset{S}{\triangle} \xrightarrow[CO]{PdCl_2} \xrightarrow[17h]{50℃} \begin{matrix}S\\S\end{matrix}$$

其他一些不使用特殊催化剂合成 1,4-二噻烷的方法如下[147~151]：

$$n\begin{Bmatrix}Br\\Br\end{Bmatrix} + n\text{Na}_2\text{S} \xrightarrow{60\sim65℃} +(\text{CH}_2\text{CH}_2\text{S})_n + 2n\text{NaBr} \xrightarrow{\text{C}_6\text{H}_5\text{OH}} \begin{matrix}S\\S\end{matrix}$$

$$\begin{Bmatrix}SH\\Br\end{Bmatrix} + \begin{Bmatrix}Br\\HS\end{Bmatrix} \xrightarrow{OH^-} \begin{matrix}S\\S\end{matrix} + 2HBr$$

$$C_2H_5SH + \triangle O \xrightarrow{C_2H_5OH,OH^-} \begin{matrix}S\\S\end{matrix}$$

$$\begin{Bmatrix}SH\\SH\end{Bmatrix} \xrightarrow{\gamma射线} \begin{matrix}S\\S\end{matrix}$$

$$\begin{Bmatrix}X\\X\end{Bmatrix} + Na + S \xrightarrow{液氨} \begin{matrix}S\\S\end{matrix}$$

二、2,5-二羟基-1,4-二噻烷

2,5-二羟基-1,4-二噻烷(2,5-dihydroxy-1,4-dithiane,FEMA 号 3826)是巯基乙醛的二聚体,具有烤肉、烤面包、土豆香、肉汤、鸡肉、蛋类、牛肉等香味特征[152],可用于蛋、咖啡、烤牛肉、猪肉、鸡肉、金枪鱼、蛤肉、河虾、牡蛎等制品、西红柿制品、芥末等食品香精[152]。

2,5-二羟基-1,4-二噻烷可以用氯乙醛与硫氢化钠反应制备[153]。

$$\underset{Cl}{\overset{O\ H}{\diagup\!\!\!\diagdown}} + NaSH \longrightarrow \underset{SH}{\overset{O\ H}{\diagup\!\!\!\diagdown}} \longrightarrow HO\begin{matrix}S\\S\end{matrix}OH$$

三、2,5-二甲基-2,5-二羟基-1,4-二噻烷

2,5-二甲基-2,5-二羟基-1,4-二噻烷(2,5-dimethyl-2,5-dihydroxy-1,4-dithiane,FEMA 号 3450)是 α-巯基丙酮的二聚体,具有鸡肉、烤香、洋葱、猪肉汤、鸡汤等香味特征[154],可用于肉类、鸡肉、蔬菜、宠物食品等香精[155]。

2,5-二甲基-2,5-二羟基-1,4-二噻烷可以用氯丙酮与硫氢化钠反应制备[156,157]。

$$\underset{Cl}{\overset{O}{\diagup\!\!\!\diagdown}} + NaSH \longrightarrow \underset{SH}{\overset{O}{\diagup\!\!\!\diagdown}} \longrightarrow HO\begin{matrix}S\\S\end{matrix}OH$$

第十七节 其他硫醚类香料

一、α-甲基-β-羟基丙基 α'-甲基-β'-巯基丙基硫醚

α-甲基-β-羟基丙基 α'-甲基-β'-巯基丙基硫醚(α-methyl-β-mercapto propyl α'-methyl-β'-hydroxyl propyl sulfide,FEMA 号 3509),商品名"966",为浅黄色液体,在 1% 时具有烤香、肉香和硫化物样香气及脂肪、肉汤香气。在 2.5mg/kg 的味觉特征是硫化物样香气、烤香、肉香、焦香及菜肴香[158,159]。可用于肉、咖啡、坚果、洋葱和大蒜等香精,在肉制品和焙烤食品中的推荐用量是 0.75mg/kg[160]。

α-甲基-β-羟基丙基 α'-甲基-β'-巯基丙基硫醚可以用下述方法制备:

(1) 2,3-环硫丁烷和 3-巯基-2-丁醇反应[158]

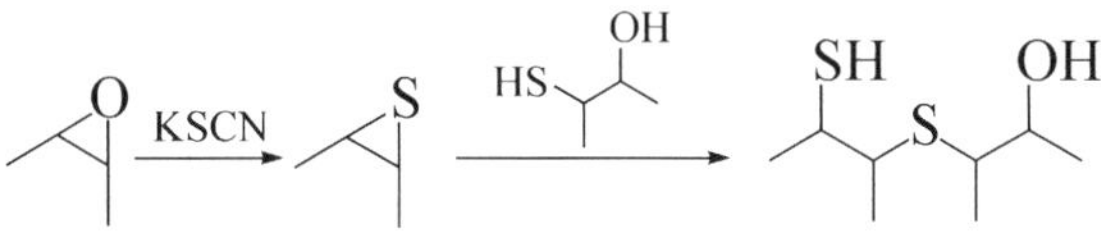

(2) 2,3-环氧丁烷和 2,3-丁二硫醇反应

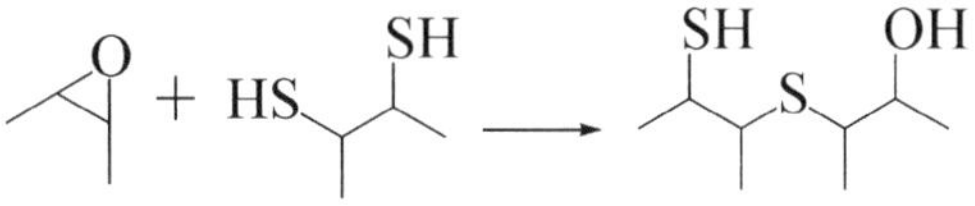

二、咖啡呋喃

咖啡呋喃(kahweofuran)天然存在于猪肉[161]、咖啡[162]的挥发性香成分中,具有肉香、咖啡、硫化物、烤香、烟熏香气,香气特征性很强[163]。

咖啡呋喃可以通过下面的路线制备[162]:

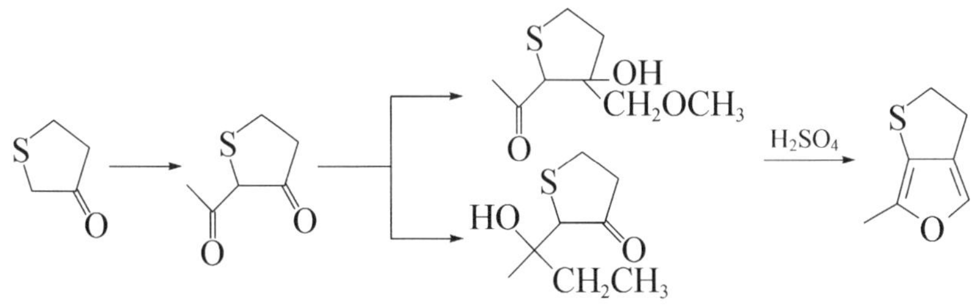

三、5-甲基-2-甲硫基呋喃

5-甲基-2-甲硫基呋喃(5-methyl-2-thiomethylfuran,FEMA 号 3366)为无色至浅黄色液体,天然存在于咖啡、麦芽中,具有洋葱、大蒜、芥末、辣根、胡椒香气以及

芥末、洋葱、大蒜、蘑菇、菜肴、咖啡味道，可用于蘑菇、调味品、咖啡、芥末、肉类、烘烤食品等香精。

5-甲基-2-甲硫基呋喃可以由 5-甲基-2-巯基呋喃与碘甲烷反应制备[10]。

$$\text{5-甲基-2-巯基呋喃} + CH_3I \xrightarrow{NaOH} \text{5-甲基-2-甲硫基呋喃} + HI$$

其他一些硫醚香料的化学结构和香味特征见表 4-13。

表 4-13 其他一些硫醚香料的化学结构和香味特征

名称	化学结构式	香味特征与应用建议
甲硫基乙酸甲酯 （FEMA 号 4003）	（结构式）	可用于焙烤食品、软饮料、糖果、肉制品等香精[111]
2,4,6-三甲基-1,3,5-三硫杂环己烷	（结构式）	壤香，坚果香，肉香[163,164]
2,2,4,4,6,6-六甲基-1,3,5-三硫杂环己烷	（结构式）	肉香[164]

参 考 文 献

[1] Landini D, Rolla F. A convenient synthesis of primary and secondary dialkyl and aryl alkyl sulfides in the presence of phase-transfer catalysts. Synthesis. 1974, 565~566

[2] Evers W J, Heinsohn H R, Mayers B J. Foodstuff flavor compositions comprising 3-furyl alkyl sulfides and processes: US, 4031256. 1977.6.21

[3] Herriott W. The phase-transfer synthesis of sulfides and dithioacetals. Synthesis. 1975, (7): 447~448

[4] Windus W, Shildneck P R. β-Hydroxyethyl methyl sulfide. Org. Syn., 1943, Col. 2: 345~417

[5] Kleipool R J. 4-Methyl-4-furfurylthio-pentanone-2: US, 4271082. 1981.6.2

[6] Kumar P, Pandey R K, Hegde V R. Anti-Markovnikov addition of thiols across double bonds catalyzed by H-Rho-zeolite. SYNLETT, 1999, (12): 1921~1922

[7] 湖上国雄. 4-アルキルチオ-2-メチルフラン类及びその利用: 日本, 公开特许公报. 昭 58-121286. 1983.7.19

[8] Kondo T, Uenoyama S, Fujita K et al. First transition-metal complex catalyzed addition of organo disulfides to alkenes enables the rapid synthesis of vicinal-dithioethers. J. Am. Chem. Soc., 1999, 121(2): 482~483

[9] Zhan Z, Lu G, Zhang Y. A novel synthesis of ally sulfides and allyl selenides via Sm-BiCl$_3$ system in aqueous media. J. Chem. Res., Synop. 1999, (4): 280~281

[10] 孙宝国, 何坚. 香料化学与工艺学. 第二版. 北京: 化学工业出版社, 2004. 490~500

[11] Tangestaninejad S, Mirkhani V. Iron(Ⅲ) 5,10,15,20-tetrakis (pentafluorophenyl)-prophyrin as an efficient catalyst for conversion of epoxides to thiiranes. J. Chem. Res., Synop. 1999, (6): 370~371

[12] Mohammadpoor-Baltork I, Khosropoor A R. Antimony trichloride: a mild, efficient and convenient catalyst for conversion of oxiranes to thiiranes. Indian J. Chem., 1999, 38B(5): 605~606

[13] Mosciano G et al. Organoleptic characteristics of flavor materials. Perfumer & Flavorist, 1998, 23(1): 33~36

[14] Mosciano G et al. Organoleptic characteristics of flavor materials. Perfumer & Flavorist, 1998, 23(5): 49~52

[15] 孙宝国, 刘玉平. 食用香料手册. 北京: 中国石化出版社, 2004. 400~401

[16] Mosciano G et al. Organoleptic characteristics of flavor materials. Perfumer & Flavorist, 1997, 22(3): 47~50

[17] Mosciano G, Sadural S, Fasano M et al. Organoleptic characteristics of flavor materials. Perfumer & Flavorist, 1990, 15(1): 19~25

[18] Mosciano G, Sadural S, Fasano M et al. Organoleptic characteristics of flavor materials. Perfumer & Flavorist, 1990, 15(2): 69~73

[19] Gianturco M A, Giammarino A S, Friedel P et al. The volatile constituents of coffee-Ⅳ Furanic and pyrrolic compounds. Tetrahedron, 1964, 20(12): 2951~2961

[20] Stoll M, Winter M, Gautschi et al. 68. Recherches sur les aromes. Helv. Chim. Acta, 1967, 50(2): 628~694

[21] Zanetti J E. α-Furfuryl ethers. J. Am. Chem. Soc., 1927, 49(1): 1065

[22] Kirner W R. α-Furfuryl chloride and its derivatives. Ⅰ. Preparation and properties of α-furfuryl chloride and a few α—furfuryl ethers. J. Am. Chem. Soc., 1928, 50(2): 1958

[23] Vernin G, Zhamkotsian R M, Metzger J. 相转移催化在芳香分子合成方面的应用. 香料与香精, 1983, (4): 17~45

[24] Newberne P, Smith R L, Doull J et al. GRAS Flavoring substances 19. Food Technology, 2000, 54(6): 66~84

[25] Werkhoff P, Brüning J, Emberger R et al. Flavor chemistry of meat volatiles: new results on flavor components from beef, pork, and chicken//Hopp R, Mori K. Recent Developments in Flavor and Fragrance Chemistry. New York: VCH Publisher, 1992. 183~213

[26] Mottram D S. Meat//Maarse H. Volatile Compounds in Foods and Beverages. Zeist, The Netherlands: TNO-CIVO Food Analysis Institute, 1991. 137~139

[27] Güntert M, Bertram H J, Emberger R et al. Thermal degradation of thiamin (Vitamin

[28] Piggott J R, Paterson A. Understanding Natural Flavor. New York: Blackie Academic & Professional, 1994. 157~159

[29] Alvarez-Ibarra C, Ouiroga M L, Toledano E. Synthesis of polysubstituted 3-thiofurans by regiospecific momo-*ipso*-substitution and *ortho*-metallation from 3,4-dibromofuran. Tetrahedron, 1996, 52(11): 4065~4078

[30] 孙宝国,郑福平,田红玉等. 一种 2-甲基-3-甲硫基呋喃的制备方法:中国,ZL96106592.3. 1997.1.29

[31] 孙宝国,郑福平.食品香味添加剂 2-甲基-3-甲硫基呋喃及 2-甲基-4-甲硫基呋喃合成的研究//陈声宗,曹声春,梅焕谋,胡艾希. 第五届全国高校化工工艺学术会论文集—化工研究进展. 北京:化学工业出版社,1996. 164~165

[32] Werkhoff P, Brüning J, Emberger R et al. Flavor chemistry of meat volatiles: new results on flavor components from beef, pork, and chicken//Hopp R, Mori K. Recent Developments in Flavor and Fragrance Chemistry. New York: VCH Publisher, 1992. 183~213

[33] Evers W J, Heinsohn H R, Mayers B J. Processes for producing 3-thia furans and 3-furan thiols: US, 3922288. 1975.11.25

[34] 黄荣初,王兴凤. 肉类香味的合成香料. 有机化学,1983,(3):175~179

[35] Evers W J, Heinsohn H R, Vock M H et al. Flavoring with (allyl)(2-methyl-3-furyl) sulfide: US, 4007287. 1977.2.8

[36] Eugene E, Tamelen V. The formation and ring-opening of alkene sulfides. J. Am. Chem. Soc., 1951, 73: 3444~3448

[37] Stephens J R, Hydock J J, Kleinholz M P. Certain *n*-dodecyl sulfides. J. Am. Chem. Soc., 1951, 73: 4050

[38] Albanese D, Landini D, Penso M. Tetrabutylammonium fluoride: A Powerful catalyst for the regioselective opening of epoxides with thiols. Synthesis, 1994, (1): 34~36

[39] Choi J, Yoon N M. Direct synthesis of thiols from halides and epoxides using hydrosulfide exchange resin in methanol. Synthesis, 1995, (4):373~375

[40] Maiti A K, Gautam K B, Prantosh B. Montmorillonite clay-catalysed regioselective ring opening of oxiranes with thiophenol: a simple synthesis of β-hydroxy sulfides. J. Chem. Research(S), 1993, (3): 325~328

[41] Iida T, Yamamoto N, Sasai H et al. New asymmetric reactions using a gallium complex: A highly enantioselective ring opening of epoxides with thiols catalyzed by a gallium-lithium-bis(binaphthoxide) complex. J. Am. Chem. Soc., 1997, 119: 4783~4784

[42] Iranpoor N, Baltork I M, Zardaloo F S. Ceric ammonium nitrate, an efficient catalyst for mild and selective opening of epoxides in the presence of water, thiols and acetic acid. Tetrahedron, 1991, 47(47): 9861~9866

[43] Maiti A K, Bhattacharyya P. Polyethylene glycol (PEG) 4000 catalysed regioselective nucleophilic ring opening of oxiranes—A new and convenient synthesis of β-hydroxy sulfone and β-hydroxy sulfide. Tetrahedron, 1994, 50(35): 10483~10490

[44] Wu J, Hou X L, Li X et al. Enantioselective ring opening of meso-epoxides with thiols catalyzed by a chiral (salen) Ti(IV) complex. Tetrahedron: Asymmetry, 1998, (9): 3431~3436

[45] Posner G H, Rogers D Z. Organic reactions at alumina surfaces. Mild and selective opening of epoxides by alcohols, thiols, benzeneselenol, amines, and acetic acid. J. Amer. Chem. Soc., 1977, (11): 8209~8210

[46] Posner G H, Rogers D Z, Kinzig C M et al. Organic reactions at alumina surfaces. Displacement reactions effected by alcohols, thiols, and acetic acid on dehydrated alumina. Tetrahedron Letters, 1975, (42): 3597~3600

[47] Takeuchi H, Kitajima K, Yamamoto Y et al. The use of proton-exchanged X-type zeolite in catalyzing ring-opening reactions of 2-substituted epoxides with nucleophiles and its effect on regioselectivity. J. Chem. Soc., Perkin Trans. 2, 1993, (2): 199~203

[48] Fringuelli F, Pizzo F, Tortoioli S et al. Thiolysis of alkyl- and aryl-1,2-epoxides in water catalyzed by $InCl_3$. Adv. Synth. Catal., 2002, 344(3+4): 379~383

[49] Vougioukas A E, Kagan H B. Oxirane ring opening reactions with thoils catalyzed by lanthanide complexes. Tetrahedron Letters, 1987, 28(48): 6065~6068

[50] Iqbal J, Pandey A, Shukla A et al. Cobalt mediated regionselective ring opening of oxiranes with benzenethiol: A mechanistic study. Tetrahedron, 1990, 46(18): 6423~6432

[51] Ian W J, Still L, Martyn J P. The generation of (samaraium) thiolates from arylthio-cyanates and their reaction with epoxides: A route to β-hydroxy sulfides. Syn. Commn., 1998, 28(5): 913~923

[52] Yamashita H, Mukaiyama T. Asymmetric ring-opening of cyclohexene oxide with various thiols catalyzed by zinc Lg-tartrate. Chem. Lett., 1985, (11): 1643~1646

[53] Yamashita H. Metal(II) d-tartrates catalyzed asymmetric ring opening of oxiranes with various nucleophiles. Bull. Chem. Soc. Jpn., 1988, 61: 1213~1220

[54] Anderson J E. Production of 2-(ethylthio)ethanol: US, 3487113. 1969. 12. 30

[55] Doumani T F. Preparations of hydroxy thio-ethers: US, 2776977. 1957. 1. 8

[56] 黄明泉,田红玉,孙宝国. α-烷硫基-2-丁酮类化合物的合成. 精细化工, 2005, 22(2): 127~129

[57] Tanabe Y, Mori K, Yoshida Y. Mild, effective and regioselective ring-opening of oxiranes using several thiosilanes promoted by tetrabutylammonium fluorides as catalys. J. Chem. Soc., Perkin Trans. 1, 1997, (5): 671~675

[58] Martin G, Sauleau J, David M et al. Hydroxythioethers ethyleneiques: synthese et rearrangement spontane. Can. J. Chem., 1992, 70: 2190~2196

[59] Brittain J, Gareau Y. Triphenylsilanethiol: A solid H_2S equivalent in the ring opening of

epoxides. Tetrahedron Letters, 1993, 34(21): 3363~3366

[60] Guindon Y, Young R N, Frenette R. Synthesis of β-trimethylsilyloxythioethers and β-hydroxythioethers by the reaction of epoxides with aryl- and alkylthiotrimethylsilanes. Syn. Commn., 11(5): 391~398

[61] Pasto D J, Cumbo C C, Balasubramaniyan P. Transfer reactions involveing boron. VI. Physical characterization and chemical properties of aryl- and alkylthioboranes. J. Am. Chem. Soc., 1966, (5): 2187~2194

[62] Pasto D J, Cumbo C C, Fraser J. Transfer reactions involveing boron. VII. The stereochemistry of ether cleavages of epoxides with phenylthioborane. J. Am. Chem. Soc., 1966, (5): 2194~2200

[63] Rawal V H, Akiba M, Cava M P. α-Methoxythioanisole. Further uses as an umpolung reagent. Synthetic Commn., 1984, 14(12): 1129~1139

[64] Guigne A, Metzner P. Ambidoselectivity of the enethiolates reaction with electrophiles: The case of epoxides. Phosphorus and Sulfur, 1985, 25: 97~102

[65] Reglier M, Julia S A. Stereoselective conversion of lithiated benzylic or allylic 3-methyl-1 (Z), 3-butadienyl sulfides into cis-disubstituted cyclopropane compounds. Tetrahedron Letters, 1983, 24(23): 2387~2390

[66] Hosomi A, Ogata K, Hoashi K et al. Novel arylthiomethylation of carbonyl compounds using arylthiomethyltrimethylsilanes catalyzed by fluorides ions. New route to β-hydroxyarylsulfides. Chem. Pharm. Bull., 1988, 36(9): 3736~3738

[67] Kitteringham J, Mitchell M B. Fluoride ion induced reaction of phenylthiomethyl - trimethylsilane(1) with aldehydes and ketones: A simple procedure for the formation of β-hydroxyphenylsulphides. Tetrahedron Letters, 1988, 29(27): 3319~3322

[68] Binns M R, Haynes R K. Hexamethylphosphoramide-mediated conjugate addition of (alkylthio)-, (phenylthio)-, and (phenylseleno)allyllithium reagents to 2-cyclopentenone. J. Org. Chem., 1981, 46: 3790~3795

[69] Yamamoto Y, Yatagai H, Maruyama K. Regioselective carbon-carbon bond formation at the α-position of a sulfer stabilized allyl carbanion via an alkylthioallylboron "ate" complex. Chem. Letters, 1979, (20): 385~386

[70] Anciaux A, Eman A, Dumont W et al. Reaction of mixed sulfo-seleno acetals and oxo-seleno acetals with n-butyllithium: A new route of substituted epoxides(1). Tetrahedron Letters, 1975, 19: 1617~1620

[71] Takeda T, Furukawa H, Fujiwara T. A convenient method for the preparation of 1-(methylthio) and 1-(phenylthio)vinyllithium reagents. Chem. Lett., 1982, (11): 593~596

[72] Shanklin J R, Johnson Cl R, Ollinger J et al. Conversion of ketones to epoxides via β-hydroxy sulfides. J. Amer. Chem. Soc., 1973, 95(10): 3429~3431

[73] Tanis S P, McMills M C, Herrinton P M. A convenient synthesis of vinyl spiro epoxides

from α,β-unsaturated ketones. J. Org. Chem., 1985, 50: 5889~5892

[74] Peterson D J. Preparation and reactions of some alkylthiomethyllithium compounds. J. Org. Chem., 1967, 32(6): 1717~1720

[75] Evers W J, Heinsohn H H, Vock M H. Uses of α-oxy(oxo) sulfides and ethers in foodstuffs and flavors for foodstuffs: US, 4044164. 1977.8.23

[76] Evers W J, Heinsohn H H. α-Oxy(Oxo) sulfides and ethers: US, 4045491. 1977.8.30

[77] Brownbridge P, Warren S. Scope and limitations of allyl sulphides synthesis by [1,2] and [1,3] phenylthio migration. J. Chem. Soc., Perkin I, 1977, (20): 2272~2282

[78] Li A H, Dai L X, Hou X L et al. Preparation of enantiomerically enriched (2R,3R)- or (2S,3S)-trans-2,3-diaryloxiranes via camphor-derived sulfonium ylides. J. Org. Chem., 1996, 61(2): 489~493

[79] Hung S M, Lee D S, Yang T K. New chiral auxiliary: Optically active thiol derived from camphor. Tetrahedron, 1990, 1(12): 873~876

[80] Carreno M C, Dominguez E, Ruano J L G et al. Studies on the stereoselectivity of hydride reducetions on 2-(methylthio)- and 2-(methylsulfonyl) cyclohecanones. J. Org. Chem., 1987, 52(16): 3619~3625

[81] Fernandez I, Llera J M, Zorrilla F et al. The fixing role of the tert-butyl group in the conformational properties of acyclic sulphur compounds. Synthesis and conformational analysis of 2-tert-butylthioderivatives of 1-phenylethanol and their o-methyl analogs. Tetrahedron, 1989, 45(9): 2703~2718

[82] Shimagaki M, Maeda T, Matsuzaki Y et al. Highly stereoselective reduction of α-methylthio and α-phenylthio ketones--synthesis of syn- and anti-β-methylthio- and β-phenylthioalcohols. Tetrahedron Letters, 1984, 25(42): 4775~4778

[83] Enders D, Piva O, Burkamp F. Aaymmetric synthesis of vicinal thioether alcohols by disastereoselective 1,2-addition of carbon nucleophiles to enantiomerically enriched α-sulfenylated aldehydes. Tetrahedron, 1996, 52(8): 2893~2908

[84] Watanabe M, Komota M, Masamichi N et al. Catalytic enantio- and disastereo- selective addition of diethylzinc to racemic α-thio- and α-seleno-aldehydes: Asymmetric synthesis of optically active vicinal thio- and seleno-alcohols. J. Chem. Soc., Perkin trans. 1, 1993, (18): 2193~2196

[85] Iriuchijima S, Kojima N. Comparative asymmetric reduction of phenylthioacetone, (±)-phenylsulfinylacetone and phenylsulfonylacetone with fermenting bakers' yeast. Agric. Biol. Chem., 1978, 42(2): 451~455

[86] Fujisawa T, Yamanaka K, Mobele B I et al. Highly enantio- and disastereoselective reduction of sulfur-functionalized cyclic ketones with bakers' yeast. Tetrhedron Letters, 1991, 32(3): 399~400

[87] Crumbie R L, Deol B S, Jacques E N et al. Asymmetric reduction of carbonyl compounds by yeast. IV Preparation of optically active β-hydroxy sulfides, sulfoxides and sulfones.

Aust. J. Chem., 1978, 31: 1965~1980

[88] Shimagaki M, Takubo H, Oishi T. Highly stereoselective reduction of α-methylthio aldehydes with allyltriphenylstannane: Synthesis of anti-β-methylthio alcohols. Tetrahedron Letters, 1985, 26 (50): 6235~6238

[89] Fleischmann C, Zbiral E. Zur umsetzung von trimethylsilyl-methylendimethyl sulfuran mit carbonylverbindungen. Tetrahedron, 1978, 34: 317~323

[90] Bury A, Earl H A, Stirling C J M. Elimination and addition reactions. Part 44. Eliminative fission of cycloalkanols. J. Chem. Soc., Perkin Trans. II, 1987, (9): 1281~1287

[91] Reisner D B. Sulfur-containing amino acids. J. Am. Chem. Soc., 1955, 78: 2132~2135

[92] Park H S, Kim S H, Park M Y et al. Facile synthesis of enol ethers by cleavage of α-bromoacetals and αbromoketals mediated by SmI_2. Tetrahedron Letters, 2001, 42: 3729~3732

[93] Nakayama J, Sugiura H, Shiotsuki A et al. Benzyne-induced fragmentation of 1,3-oxathiolanes. A novel method for deprotection of carbonyl groups, preparation of phenyl vinyl sulfides, and 1,2-carbonyl transposition. Tetrahedron Letters, 1985, 26(18): 2195~2198

[94] Yaday V K, Fallis A G. Cyclopentane synthesis and annulation: Intramolecular radical cyclization of acetals. Tetrahedron Letters, 1988, 29(8): 897~900

[95] Leffetter B E, Brown R K. The reductive cleavage of 1,3-oxathiolanes with lithium aluminum hydride in the presence of aluminum chloride or boron trifluoride. Can. J. Chem., 1963, 41: 2671~2682

[96] Yadav V, Fallis A G. Free-radical based cycloalkanol synthesis and annulation from thioacetal precursors. Can. J. Chem., 1991, 69: 779~789

[97] Khurana J M, Sahoo P K. Chemoselective alkylation of thiols: A detailed investigation of reactions of thiols with halides. Syn. Commun., 1992, 22(12): 1691~1702

[98] Angeletti E, Tundo P, Venturello P. Gas-liquid phase-transfer synthesis of phenyl ethers and sulphides with carbonate as base and carbowax as catalyst. J. Chem. Soc., Perkin I, 1982, (5): 1137~1141

[99] Tundo P, Giovanni M, Trotta F. Gas-liquid phase-transfer catalysis: A new continuous-flow method in organic synthesis. Ind. Eng. Chem. Res., 1989, 28: 881~890

[100] Strege P E. Inorganic halide salt catalysts for hydroxyalkylation of phenols or thiophenols: US, 4341905. 1982.7.27

[101] Fitt P S, Owen L N. Dithiols. Part XX. The occurrence of transthiomethylation in some reactions of 2:3-bismethylthiopropanol and related compounds. J. Chem. Soc., 1957, (7): 2250~2256

[102] Kondo T, Morisaki Y, Uenoyama S Y et al. First ruthenium-catalyzed allylation of thiols enables the general synthesis of allylic sulfides. J. Am. Chem. Soc., 1999, 121(37): 8657~8658

[103] Yamauchi K, Nakamura K, Kinoshita M. Trimethylselenonium hydroxide: A new methylating agent. Tetrahedron Letters, 1979, (20): 1787~1790

[104] Labat Y. New hydroxy-thia-alkenes, and processes for their preparation and use: US, 4847420. 1989.7.11

[105] Kharasch M S, Nudenberg W, Mantell G J. Reactions of atoms and free radicals in solution. XXV. The reactions of olefins with mercaptans in the presence of oxygen. J. Org. Chem., 1950, 16: 524~532

[106] 李彬城. 脱氧核苷磷酸基团保护剂的合成与磷酸基团的保护研究. 化学试剂, 1983, 5(1): 50~51

[107] Carretero J C, Garcia-Ruano J L, Carmen M et al. Rodriguez. Syntheses of β-heterosubstituted thioethers. J. Chem. Research(S), 1985, (1): 6~7

[108] Hurd C D, Wilkinson K. Reaction of chloroalkyl sulfides with sodium. J. Am. Chem. Soc., 1947, 71: 3429~3433

[109] Dawson T P. The conversion of certain mercaptans into acetates and sulfides. J. Am. Chem. Soc., 1947, 69: 1211~1212

[110] Baldwin A W, Piggott H A. Solubilized higher aliphatic sulphides and process for their production: US, 2100297. 1937.11.23

[111] Smith R L, Doull J, Feron V J et al. GRAS flavoring substances 20. Food Technology. 2001, 55(12): 34~55

[112] 孙宝国, 田红玉, 黄明泉等. 一类1-烷硫基-2-甲氧基丁烷类化合物及其制备方法: 中国, 200510082705.3. 2006.1.18

[113] Goldsworthy L J, Harding G F, Norris W L et al. Some sulphides containing the 2-chloroethyl group. J. Chem. Soc., 1948, (4): 2177~2179

[114] Eliel E L, Doyle T W, Daignault R A et al. A facile synthesis of β-alkoxy mercaptans and β-thioalkoxy mercaptans. J. Am. Chem. Soc., 1966, 88(8): 1828~1830

[115] Newman B C, Eliel E L. Reduction with metal-ammonia combinations. Ⅲ. Synthesis of β- and γ-alkylthiomercaptans from 1,3-dithiolanes and 1,3-dithianes. J. Org. Chem., 1970, 35(11): 3641~3646

[116] Brown E D, Iqbal S M, Owen L N. The reductive fission of methyl sulphides, 1,3-dithiolans, and a 1,3-oxathiolan by sodium in liquid ammonia. J. Chem. Soc. (C), 1966, 415~419

[117] Kikugawa Y. Chemistry of amine-boranes. Part 10. Synthesis of sulphides from dithioacetals using pyridine-borane in acid. J. Chem. Soc., Perkin Trans. I, 1984, (4): 609~610

[118] Huxley E E. Purification of mercaptoalkyl n-alkyl sulfides: US, 3786101. 1974.12.18

[119] Snyder H R, Stewart J M. Beta mercapto thioethers: US, 2490984. 1949.12.13

[120] Takeuchi H, Nakajima Y. Novel $S_N 2$ ring-opening reactions of 2- and 2,2-substituted thiiranes with thiols using Na^+-exchanged X-type zeolite or triethylamine in methanol.

J. Chem. Soc., Perkin Trans. 2, 1998, (11): 2441~2446

[121] Culvenor C C J, Davies W, Heath N S. The preparation and reactions of aliphatic and alicyclic ethylene sulphides. J. Chem. Soc., 1949, (1): 282~287

[122] Meade E M, Woodward F N. Some reactions of ethylene sulphide, and a new method of preparation of vicinal dithiols. J. Chem. Soc., 1948, (4): 1894~1895

[123] Helmkamp G K, Olsen B A, Koskinen J R. Stereochemistry of the addition of dialkylalkylthiosulfonium salts to alkenes and alkynes. J. Org. Chem., 1965, 30(5): 1623~1626

[124] Bresson C R, Dix J S. Production of dithiols from aziridines: US, 3424800. 1969.1.28

[125] Dix J S, Bresson C R. Nucleophilic displacement reactions of β-amino mercaptans. J. Org. Chem., 1967, 32(2): 282~285

[126] Newberne P, Smith R L, Doull J et al. GRAS flavoring substances 18. the 18th publication by the Flavor and Extract Manufactures' Association's Expert Panel on recent progress in the consideration of flavoring ingredients generally recognized as safe under the food additives amendment. Food Technology, 1998, 52(9): 79~92

[127] Bradsher C K, Brown F C, Grantham R J. Synthesis and fungistatic activity of some 5-(1-Methyl-2-thioalkylethylidene)-rhodanines. J. Am. Chem. Soc., 1954, 76: 114~115

[128] Morey G H. Acetonyl thio ethers: US, 2363462. 1944.11.21

[129] 刘玉平,孙宝国,郑福平等. 3-甲硫基-1-丙醇的合成研究. 香料香精化妆品, 2003, (6): 1~2

[130] David B. Reisner D B. Sulfur-containing amino acids. J. Am. Chem. Soc. 1956, 78(3): 2132~2135

[131] 高振永,柯友之,王亨权等. 合成食品香味增效剂 α-羰基硫醇和 α,β-二羰基硫醚的研究. 北京轻工业学院学报, 1985, 3(1): 46~56

[132] Mosciano G. Organoleptic characteristics of flavor materials. Perfumer & Flavorist, 1996, 21(4): 51~55

[133] 孙宝国,郑福平. 3-甲硫基丙酸甲酯的合成研究. 北京轻工业学院学报, 1995, 13(2): 46~48

[134] Hurd C D, Gershbein L L. Reactions of mercaptans with acrylic and methacrylic derivatives. J. Am. Chem. Soc., 1947, 69(3): 2328~2335

[135] Oser B L, Ford R A, Bernard B K. Recent progress in the consideration of flavoring ingredients under the food additives amendment. 13. GRAS substances. Food Technology, 1984, 38(10): 66~89

[136] Ray P C, Bose-Ray K C. Lengthened chain compounds of sulfur. Quart J. Indian Chem. Soc, 1926, 3: 75~80

[137] Tucker N B, Reid E E. Cyclic and polymeric compounds from the reaction of ethylene mercaptan with polymethylene halides. J. Am. Chem. Soc., 1933, 55(1): 775~781

[138] Fuson R C, Ziegler J B. 2-Chloroethyl 2-hydroxyethyl sulfide. J. Org. Chem., 1946,

55(11): 510~512
[139] Swistak E. Action of cation exchangers on some Ddiols. Compt. Rend, 1955, 240: 1544~1545
[140] Yur'ev Y K, Novitskii K V. Dehydration of thiodiethyleneglycol and disproportionation of p-oxathiane. Doklady Akad Nauk S S S R, 1949, 68: 717~719
[141] Yur'ev Y K, Novitskii K V. Reaction of ethylene oxide with hydrogen sulfide in the presence of alumina. Doklady Akad Nauk S S S R, 1948, 67: 285~288
[142] Yur'ev Y K, Novitskii K V. α-Oxides and synthesis of compounds of the thiophone series. Zhur Obshchei Khim, 1952, 22: 2187~2189
[143] Yur'ev Y K, German L S. Catalytic transformations of ethylene sulfide and ethanedithiol. Zhur Obshchei Khim, 1955, 25: 2527~2529
[144] Eisfeld Wolfgang T, Weil Edward D. Synthesis of p-dithiane: US, 3503991. 1970.3.31
[145] Warner P F. p-Dithiane: US, 3070605. 1962.12.15
[146] Scheben J A, Mador I L. Cyclodimerizing oxirines, thiiranes, and aziridines: US 3480632. 1969.11.25
[147] Fuson R C, Lipscomb R D, Mckusick B C et al. Thermal conversion of mustard gas to 1,2-bis(2-chloroethylthio)ethane and bis[2-(2-chloroethylthio) ethyl] sulfide. J. Org. Chem., 1946, 11: 513~517
[148] 陈海涛,孙宝国,梁梦兰等.1,4-二噻烷类香料化合物的合成及其在食品中的应用.北京工商大学学报(自然科学版),2003,21(2):1~3
[149] Warner P F, Lambert C E. Preparation of p-Dithiane and diethyl sulfide: US, 3119838. 1964.1.28
[150] Chirakadze G G, Mosashvili G A. Formation of dithianes during irradiation of alkanedithiols. Tezisy Dokl Nauchn Sess Khim Tekhnol Org. Soedin Sery Sernistykh Neftei, 13th. 1974. 281~286
[151] Chorbadzhiev S, Markov P. Interaction of haloorganic compounds with sodium sulfides in liquid ammonia. Dokl bolg Akad Nauk, 1972, 25(6): 763~766
[152] Mosciano G. The creative flavorist. Perfumer & Flavorist, 1999, 24 (6): 10~13
[153] Hromatka O, Haberl R. 2,5-Dihydroxy-1,4-dithiane. Monatsh, 1954, 85:1088~1096
[154] Rowe D J. Aroma chemicals for savory flavors. Perfumer & Flavorist, 1998, 23 (4): 9~16
[155] Shaikh Y. 肉味香精中的合成香料. 香料香精化妆品, 1985, 1 (3): 53~56
[156] June M C, Chritopher G. Dialkyl dihydroxy dithianes as flavoring agents: DE, 2426865. 1975.1.2
[157] Russell G A et al. Aliphatic semidiones 44. Spin probes derived from dithiols. J. Am. Chem. Soc., 1985, 107(14): 4175~4182
[158] Wilson R A, Giacino C. Flavoring with sulfur containing compounds: US, 3892878. 1975.7.1

[159] Mosciano G, Long J, Holmgren C et al. Organoleptic characteristics of flavor materials. Perfumer & Flavorist, 1996, 21(2): 47~49

[160] Oser B L, Ford R A. Recent progress in the consideration of flavoring ingredients under the food additives amendment, 10. GRAS Substances. Food Technology, 1977, 31(1): 65~74

[161] Werkhoff P, Brüning J, Emberger R et al. Flavor chemistry of meat volatiles: new results on flavor components from beef, pork, and chicken//Hopp R, Mori K. Recent Developments in Flavor and Fragrance Chemistry. New York: VCH Publisher, 1992. 183~213

[162] Büchi G, Degen P. Structure and synthesis of kahweofuran, a constituent of coffee aroma. J. Org. Chem., 1971, 36(1): 199~200

[163] Werkhoff P, Brüning J, Emberger R et al. Studies on volatile sulphur-containing flavour components in yeast extract//Bhattacharyya S C, Sen N, Sethi K L. 11th International Congress of Essential Oils, Fragrances and Flavours. Proceedings: Volume4, Chemistry Analysis and Structure. New Delhi: Oxford & Ibh publishing Co. Pvt. Ltd., 1989. 215~243

[164] Macleod G. The flavor of beef//Shahidi F. Flavor of Meat and Meat products. New York: Blackie Academic & Professional, 1994. 4~33

第五章 二硫醚类香料

二硫醚类化合物可以用下面的结构通式表示：

$$\begin{matrix} & S & R' \\ R & S & \end{matrix} \qquad 或 \qquad RSSR'$$

式中：R、R′为烃基。R、R′相同的为对称二硫醚；R、R′不相同的为不对称二硫醚；R、R′相连的为环二硫醚。

二硫醚类化合物的化学性质稳定,广泛存在于食品香成分中,如二甲二硫醚存在于清酒、熟鸡、咖啡中[1];二烯丙基二硫醚是大蒜油的香成分[2];2,4,5-三硫杂己烷是烤肉的香成分;2,5-二甲基-1,2,4-三噻烷是牛肉香成分;2,6-二甲基-1,2,4,5-四噻烷是羊肉的香成分等。

二硫醚类化合物一般具有与食品有关的香气,如菜肴、葱蒜、烤肉等香气,是咸味食品香精的重要原料,在食品中用量一般为百万分之一级。

第一节 二硫醚的一般合成方法

一、硫醇氧化

硫醇很容易氧化生成对称二硫醚,常用的氧化剂有二氧化锰[3]、二甲基亚砜[4]、氧气(空气)[5~7]、过氧化氢[8]、水合肼[9]、碘[10]等。例如：

$$RSH + MnO_2 \longrightarrow RSSR + MnO + H_2O$$

$$\text{PhSH} + O_2 \longrightarrow \text{PhSSPh}$$

在微波作用下,用空气氧化硫醇合成对称二硫醚的反应时间大大缩短[11]。

二、卤代烷和二硫化二钠反应

卤代烷和二硫化二钠在醇溶液中加热反应是制取对称二硫醚的常用方法。例如[12~14]：

$$\text{RBr} + Na_2S_2 \longrightarrow \text{RSSR} + NaBr$$

$$\text{C}_6\text{H}_{11}\text{Cl} + \text{Na}_2\text{S}_2 \longrightarrow \text{C}_6\text{H}_{11}\text{-S-S-C}_6\text{H}_{11} + \text{NaCl}$$

三、醛和硫氢化钠反应

醛和硫氢化钠在碱溶液中加热反应可制取对称二硫醚。例如[15]：

$$\text{(furan)-CHO} + \text{NaSH} \longrightarrow \text{(furan)-CH}_2\text{SSCH}_2\text{-(furan)}$$

四、卤代烷与硫代硫酸钠反应

卤代烷与硫代硫酸钠反应先生成 Bunte 盐，然后再与硫醇反应可以制取不对称二硫醚[16]。

$$\text{RX} + \text{Na}_2\text{S}_2\text{O}_3 \longrightarrow \text{RS}_2\text{O}_3\text{Na} \xrightarrow{\text{R}'\text{SH}} \text{RSSR}'$$

五、格氏试剂和二氯氧硫反应

格氏试剂和二氯氧硫反应可以制取对称二硫醚或不对称二硫醚[17]。

$$\text{RMgCl} + \text{SOCl}_2 \longrightarrow \text{RSOCl} \xrightarrow{\text{R}'\text{SH}} \text{R-S(=O)-S-R} \xrightarrow{[\text{H}]} \text{RSSR}'$$

六、烷基硫化氯和硫醇钠反应

烷基硫化氯和硫醇钠在碱性溶液中加热反应可以制取对称二硫醚或不对称二硫醚[18]。

$$\text{RSCl} + \text{R}'\text{SNa} \xrightarrow{\text{NaOH}} \text{RSSR}' + \text{NaCl}$$

七、由硫醇锂制备

由硫醇锂可以制备对称二硫醚或不对称二硫醚[19]。

$$\text{R-CH=CH-SLi} + \text{I}_2 \longrightarrow \text{R-CH=CH-S-S-CH=CH-R}$$

$$\text{R-CH=CH-SLi} + \text{R}'\text{SCN} \longrightarrow \text{R-CH=CH-S-S-R}'$$

八、二乙基偶氮二羧酸酯与硫醇反应

二乙基偶氮二羧酸酯与硫醇反应可以制备对称二硫醚或不对称二硫醚[20]。

$$RSH + C_2H_5OOCN=NCOOC_2H_5 \longrightarrow C_2H_5OOCNH-\underset{\underset{SR}{|}}{N}COOC_2H_5$$

$$\xrightarrow{R'SH} RSSR' + C_2H_5OOCNH-NHCOOC_2H_5$$

九、烷硫基邻苯二甲酰亚胺等与硫醇反应

烷硫基邻苯二甲酰亚胺（或烷硫基琥珀酰亚胺，或烷硫基马来酰亚胺）与硫醇反应可以制备对称二硫醚或不对称二硫醚[21~23]。

十、由烷硫基三甲基硅烷合成

由烷硫基三甲基硅烷可以合成对称二硫醚或不对称二硫醚[24,25]。例如：

$$RSSi(CH_3)_3 + R'SCl \longrightarrow RSSR' + (CH_3)_3SiCl$$

$$RSSi(CH_3)_3 + R'-\overset{\overset{O}{\|}}{S}-S-R' \longrightarrow RSSR' + (CH_3)_3SiOSi(CH_3)_3$$

十一、烃二硫基甲酸酯与硫醇反应

烃二硫基甲酸酯与硫醇反应可以制备对称二硫醚或不对称二硫醚[26]。例如：

$$RSS\overset{\overset{O}{\|}}{C}-OR_1 + R_2SH \longrightarrow RSSR_2$$

十二、硫化氰与硫醇反应

硫化氰与硫醇反应可以制备对称二硫醚或不对称二硫醚[27,28]。例如：

$$RSH + (SCN)_2 \longrightarrow RSSCN \xrightarrow{R'SH} RSSR'$$

十三、硫醇与四氧化二氮反应

硫醇与四氧化二氮反应可以制备对称二硫醚或不对称二硫醚[29,30]。例如：

$$RSH + N_2O_4 \longrightarrow [R—S—N=O] \xrightarrow{R'SH} RSSR'$$

第二节 对称二硫醚类香料

一、二糠基二硫醚

二糠基二硫醚（FEMA 号 3257）是香油[31]、咖啡、炖牛肉的挥发性香成分[32,33]，香气特征为硫化物样、咖啡样、烤香和坚果香；味觉特征是在 20mg/kg 具有烤香、硫化物样、大蒜样和肉香[34]。可用于软饮料、冰淇淋、糖果、焙烤食品、口香糖、肉制品、奶制品等香精配方中，在食品中的用量一般为 3mg/kg[35,36]。

二糠基二硫醚可以糠硫醇为原料，用二甲基亚砜氧化合成[37]：

（反应式：呋喃-CH₂SH + (CH₃)₂SO ⟶ 呋喃-CH₂-SS-CH₂-呋喃）

二、二(2-甲基-3-呋喃基)二硫醚

二(2-甲基-3-呋喃基)二硫醚发现于炖牛肉和炖猪肉的挥发性香成分中，该香料具有饱满的肉香，其香气阈值 0.02ng/kg，香味阈值 2ng/kg，是已知的香气阈值最低的香料之一[32,38~40]。二(2-甲基-3-呋喃基)二硫醚是维生素 B_1 热降解产物挥发性香成分之一[41]，在用 Maillard 反应制备的鸡肉香精中也发现了该化合物[42]，它对肉香味特别是炖牛肉和烤牛肉香味的形成起主要作用[43]。该香料 FEMA 号 3259，在肉制品中的建议用量为 0.1mg/kg[36]。

2-甲基-3-巯基呋喃在稀碱溶液中用碘和碘化钾的水溶液氧化可以合成二(2-甲基-3-呋喃基)二硫醚[44]；在己烷溶液中用空气氧化 2-甲基-3-巯基呋喃也可以合成二(2-甲基-3-呋喃基)二硫醚[7]；用二甲亚砜氧化 2-甲基-3-巯基呋喃也可合成二(2-甲基-3-呋喃基)二硫醚[45]。

（反应式：2-甲基-3-巯基呋喃 + (CH₃)₂SO ⟶ 二(2-甲基-3-呋喃基)二硫醚）

一些常见的对称二硫醚化合物的化学结构和香味特征见表 5-1。

表 5-1 一些常见的对称二硫醚化合物的化学结构和香味特征

名称	化学结构式	香味特征或应用建议
二甲基二硫醚 (FEMA 号 3536)	CH₃-SS-CH₃	强烈的、洋葱、甘蓝、玉米罐头样香气和味道。可用于焙烤食品、肉制品、奶制品、糖果等香精[46]
二乙基二硫醚 (FEMA 号 4093)	C₂H₅-SS-C₂H₅	可用于焙烤食品、肉制品、奶制品、调味品等香精[47]
二丙基二硫醚 (FEMA 号 3228)	C₃H₇-SS-C₃H₇	洋葱、大蒜、青葱香气和味道。可用于汤料、调味品等香精[48]
二烯丙基二硫 (FEMA 号 2028)	CH₂=CHCH₂-SS-CH₂CH=CH₂	强烈的、洋葱、大蒜、芥末香气;洋葱、大蒜、芥末、肉香味道。可用于调味品、肉制品等香精[49]
二环己基二硫醚 (FEMA 号 3448)	C₆H₁₁-S-S-C₆H₁₁	鸡蛋、洋葱、大蒜、蛤肉、蟹肉、咖啡香气;洋葱、肉香、蛤肉、蟹肉、咖啡、可可味道[50]
二苯基二硫醚 (FEMA 号 3225)	C₆H₅-S-S-C₆H₅	葱、蒜、萝卜香味。可用于饮料、糖果、冰淇淋、果冻等香精[48]
二苄基二硫醚 (FEMA 号 3617)	C₆H₅CH₂-SS-CH₂C₆H₅	焦香、焦糖香。可用于焙烤食品、软饮料、糖果、肉制品、果冻等香精[51]

第三节 不对称二硫醚类香料

一、甲基 2-甲基-3-呋喃基二硫醚

甲基 2-甲基-3-呋喃基二硫醚(FEMA 号 3573),商品名"719",是最重要的肉味香料之一,在 1% 时具有硫化物样、肉香、蔬菜样及洋葱、烤牛肉香气。味觉特征是硫化物样、肉香、蔬菜样、肉汤及菜肴香味。可用于牛肉、洋葱、大蒜、西红柿、咖啡、焦糖和坚果香精[52]。在食品中的用量一般为 0.5mg/kg[46]。甲基 2-甲基-3-呋喃基二硫醚发现于烤牛肉、烤猪肉的挥发性香成分中,在酵母提出物和咖啡的挥发性香成分也有发现,其香气阈值 10ng/kg,香味阈值 0.1~1μg/kg[32,53]。

甲基 2-甲基-3-呋喃基二硫醚可以由二甲基二硫醚和 2-甲基-3-呋喃硫醇反应制备[54,55]。

[反应式: 2-甲基-3-呋喃硫醇 + CH₃SSCH₃ → 甲基 2-甲基-3-呋喃基二硫醚 + 二(2-甲基-3-呋喃基)二硫醚]

此方法的副产物二(2-甲基-3-呋喃基)二硫醚比较多。

采用硫代硫酸钠和与碘甲烷(或硫酸二甲酯,或碳酸二甲酯)反应生成的 Bunte 盐再与 2-甲基-3-呋喃硫醇反应合成甲基 2-甲基-3-呋喃基二硫醚的方法其副产物少[56~58]。

$$(CH_3)_2SO_4 + Na_2S_2O_3 \longrightarrow CH_3S_2O_3Na$$

[反应式: 2-甲基-3-呋喃硫醇 + CH₃S₂O₃Na --OH⁻→ 甲基 2-甲基-3-呋喃基二硫醚]

二、丙基 2-甲基-3-呋喃基二硫醚

丙基 2-甲基-3-呋喃基二硫醚(FEMA 号 3607)具有肉香和烤肉香,可用于焙烤食品、肉制品、汤类、快餐食品、坚果类食品,在食品中的用量一般为 0.5mg/kg[51]。

丙基 2-甲基-3-呋喃基二硫醚可以由二丙基二硫醚和 2-甲基-3-呋喃硫醇反应制备[54,55]。

[反应式: 2-甲基-3-呋喃硫醇 + C₃H₇SSC₃H₇ → 丙基 2-甲基-3-呋喃基二硫醚 + 二(2-甲基-3-呋喃基)二硫醚]

丙基 2-甲基-3-呋喃基二硫醚也可以采用硫代硫酸钠和与卤丙烷(氯丙烷或溴丙烷)反应生成的 Bunte 盐再与 2-甲基-3-呋喃硫醇制备[56,59]。

$$CH_3CH_2CH_2Br + Na_2S_2O_3 \longrightarrow CH_3CH_2CH_2S_2O_3Na$$

[反应式: 2-甲基-3-呋喃硫醇 + CH₃CH₂CH₂S₂O₃Na --OH⁻→ 丙基 2-甲基-3-呋喃基二硫醚]

三、甲基糠基二硫醚

甲基糠基二硫醚(FEMA 号 3362)是咖啡、白面包、猪肉[60]、香油[31]、猪肝[61]等食物的挥发性香成分,具有焦香和肉香。该香料在水中的香气阈值是 $0.04\mu g/kg$[62],可用于多种食品香精中。在软饮料、冰淇淋、肉制品等中的用量一般为 1mg/kg;在焙烤食品、糖果等中的用量一般为 2mg/kg[63]。

第五章 二硫醚类香料

甲基糠基二硫醚可以用碘甲烷、硫代硫酸钠和糠硫醇反应合成[64]，也可以用硫酸二甲酯和碳酸二甲酯与硫代硫酸钠和糠硫醇反应合成[57]。

$$CH_3I + Na_2S_2O_3 \longrightarrow CH_3S_2O_3Na$$
$$(CH_3)_2SO_4 + Na_2S_2O_3 \longrightarrow CH_3S_2O_3Na$$

一些常见的不对称二硫醚类香料的化学结构和香味特征见表5-2。

表5-2 一些常见的不对称二硫醚类香料的化学结构和香味特征

名称	化学结构式	香味特征或应用建议
甲基乙基二硫醚（FEMA号4040）		可用于焙烤食品、饮料、糖果、肉制品、奶制品等香精[65]
甲基丙基二硫醚（FEMA号3201）		葱蒜、小萝卜、芥菜、西红柿、土豆、大蒜香气；甜的、洋葱、大蒜、西红柿、土豆、葱蒜、蔬菜味道[66]
甲基戊基二硫醚（FEMA号4025）		可用于焙烤食品、软饮料、糖果、肉制品、奶制品等香精[65]
甲基异戊基二硫醚（FEMA号4168）		可用于焙烤食品、软饮料、肉制品等香精[47]
甲基烯丙基二硫醚（FEMA号3127）		洋葱香、韭菜香、大蒜及盐渍的蒜头香味。可用于肉制品、调味品等香精[48]
乙基丙基二硫醚（FEMA号4041）		可用于焙烤食品、饮料、糖果、肉制品、奶制品等香精[65]
乙基丁基二硫醚（FEMA号4027）		可用于焙烤食品、饮料、糖果、奶制品等香精[65]
烯丙基丙基二硫醚（FEMA号4073）		强烈的蒜头香味。可用于焙烤食品、肉制品、奶制品、调味品等香精[47]
丙烯基丙基二硫醚（FEMA号3227）		烹煮洋葱似的香气和味道。可用于肉制品、调味品等香精[48]
甲基苄基二硫醚（FEMA号3504）		甘蓝、蔬菜、青香香气；蔬菜、青香、洋葱味道。可用于焙烤食品、肉制品等香精[67]

续表

名称	化学结构式	香味特征或应用建议
乙基(2,5-二甲基-3-呋喃基)二硫醚	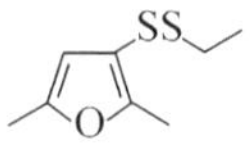	甜的、烤香香气;甜的、烤咖啡样香味[54]
丙基(2-甲基-3-呋喃基)二硫醚	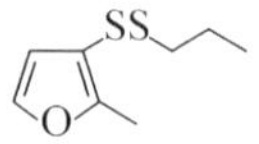	甜的、烤肉香、肉香香气;烤牛肉、烤坚果香味;带有咖啡、肝、血和金属样香韵[54]
环己基(2-甲基-3-呋喃基)二硫醚	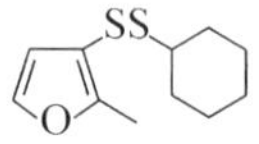	甜的、肝香、肉香香气;甜的、肝香、肉香香味[54]
异戊基(2,5-二甲基-3-呋喃基)二硫醚	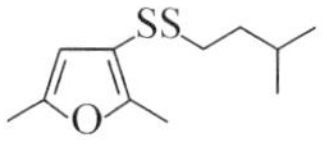	硫化物样、橡胶样、肝香气;甜的、橡胶样、硫化物样香味[54]

第四节 环二硫醚类香料

环二硫醚化合物可以通过二硫醇用亚砜氧化制备。例如[68]:

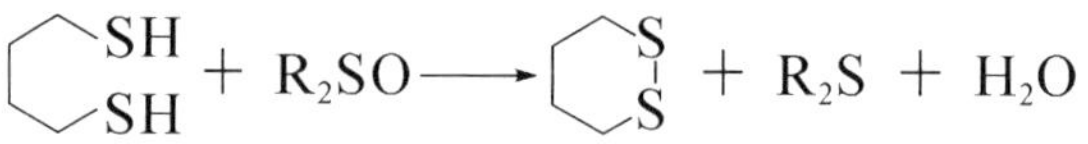

$$\ce{\underset{SH}{\overset{SH}{\bigcirc}} + R_2SO \longrightarrow \underset{S}{\overset{S}{\bigcirc}} + R_2S + H_2O}$$

一些常见环二硫醚类香料的化学结构和香味特征见表5-3。

表5-3 一些常见环二硫醚类香料的化学结构和香味特征

名称	化学结构式	香味特征
3-甲基-1,2-二硫杂环戊烷	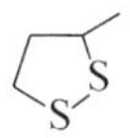	炖牛肉香、洋葱香、金属样
1,2,4-三硫杂环戊烷		烤牛肉香、硫化物样
3-甲基-1,2,4-三硫杂环戊烷	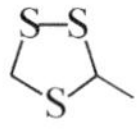	洋葱香、葱蒜香、肉香[69]
3,5-二甲基-1,2,4-三硫杂环戊烷	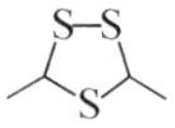	脂肪香、辛香、煮牛肉香、洋葱香、坚果香、肉香、硫化物样[69~71]
3-甲基-5-乙基-1,2,4-三硫杂环戊烷	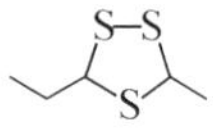	药草香、洋葱香、葱蒜香、坚果香、烤花生香[69]

续表

名称	化学结构式	香味特征
3,5-二乙基-1,2,4-三硫杂环戊烷(FEMA号 4030)		硫化物样、橡胶样、龙蒿香、小茴香香[69]
3-甲基-5-正丙基-1,2,4-三硫杂环戊烷		脂肪香、龙蒿香、小茴香香[69]
3-异丁基-5-甲基-1,2,4-三硫杂环戊烷		淡青香、可可香、猪肉香、烤肉香[70]
3,5-异丁基-1,2,4-三硫杂环戊烷		油煎培根肉香、烤猪肉皮香
3-甲基-5-正丁基-1,2,4-三硫杂环戊烷		脂肪香[69]
3-甲基-5-正戊基-1,2,4-三硫杂环戊烷		硫化物样、洋葱香、葱蒜香、肉香[69]
3-甲基-1,2,4-三硫杂环己烷		烤肉香[71,72]
2-甲基-3,4-二硫杂环己酮		血、硫化物样、肉香[70]
3,6-二甲基-1,2,4,5-四硫杂环己烷		羊肉香[73]

参 考 文 献

[1] Flament I. Molecular gastronomy. Perfumer & Flavorist, 1997, 22(1): 1~8
[2] Rowe D. More fizz for your buck: High-impact aroma chemicals. Perfumer & Flavorist, 2000, 25(5): 1~19
[3] Wallace T J. Reaction of thiols with metals. Ⅰ. Low-temperature oxidation by metal oxides. J. Org. Chem., 1966, 31: 1217~1221
[4] Yiannios C N, Karabinos J V. Oxidation of thiols with dimethyl sulfoxide. J. Org. Chem., 1963, 28: 3246~3248

[5] Hirano M, Monobe H, Yakabe S et al. Hydrotalcite clay-catalyzed air oxidation of thiols. J. Chem. Res., Synop. 1999, (6): 374~375

[6] Iranpoor N, Zeynizadeh B. Air oxidative coupling of thiols to disulfides catalyzed by(Ⅲ)/NaI. Synthesis. 1999, (1): 49~50

[7] Evers W J. Certain furan-3-thiols, certain dihydro derivatives thereof and 2,5-dimethyltetrahydrofuran-3-thiol: US, 4055578. 1977.10.25

[8] Kesavan V, Bonnet-Delpon D, Begue J P. Oxidation in fluoro alcohols: mild and efficient preparation of disulfides from thiols. Synthesis, 2000, (2): 223~225

[9] Rajaram S, Reddy G S, Iyengar D S. An unusual action of hydrazine hydrate on arylthiols: a new facile method for synthesis of disulfides. Indian J. Chem., 1999, 38B(6): 639~640

[10] Small L V D, Bailey J H, Cavallito C J. Alkyl thiolsulfinates. J. Am. Chem. Soc., 1947, 69(2): 1710~1713

[11] Sainte-Marie L, Guibe-Jampel E, Therisod M. Facile solvent free oxidation of thiols mediated by mineral supports. Tetrahedron Lett., 1998, 39(52): 6090~9662

[12] Challenger F, Rawlings A A. The formation of organo-metalloidal and similar compounds by micro-organisms. J. Chem. Soc., 1937, (1): 868~875

[13] 田红玉,孙宝国,徐理阮. 用相转移催化法合成二丙基二硫的研究. 北京轻工业学院学报, 1997, 15(1): 66~69

[14] 陆研,乔旭,汤吉海等. 香料二环己基二硫醚合成产物的鉴别与分析. 日用化学工业, 2006, 36(1): 22~25

[15] Gilman H, Hewlett A P. The vesicant action of chloro-alkyl furfuryl sulfide. J. Am. Chem. Soc., 1930, 52: 2141~2144

[16] Milligan B, Swan J M. Unsymmetrical dialkyl disulphides from Bunte salts. J. Chem. Soc., 1963, (12): 6008~6012

[17] 陈煜强. 香料用含硫化合物的开发. 香料香精化妆品, 1992, (3): 17~33

[18] Isozaki Masashi, Nanba Byolchi, Thiourea derivative: 公开特许公报, 昭 61-57510. 1944.6.3

[19] Wijers H W, Boelens H, Gen A V D. Synthesis and some properties of 1-alkenyl alkyl disulfides and di(1-alkenyl) disulfides. Recueil, 1969, 88: 519~529

[20] Mukaiyama T, Takahashi K. A conventient method for the preparation of unsymmetrical disulfides by the use of diethyl azodicarboxylate. Tetrahedron Lett., 1968, 5907~5908

[21] Harpp D N. Organic sulfur chemistry Ⅷ. A new synthesis of unsymmetrical disulfides. Tetrahedron Lett., 1970, (41): 3551~3554

[22] Boustany K S. Chemistry of sulfur compounds Ⅵ. Novel method for the preparation of disulfides. Tetrahedron Lett., 1970, (41): 3547~3549

[23] Bwhforouz M. Alkyl and aryl sulfenimides. J. Org. Chem., 1969, 34(1): 51~55

[24] Capozzi G et al. Synthesis of disulfides and trisulfides via organosilicon compounds. Gazz. Chim. Ital., 1990, 120(7): 421~426

[25] Harpp D N. Organic sulfur chemistry 29. Use of the trimethylsilyl group in synthesis preparation of sulfinate esters and unsymmetrical disufides. J. Org. Chem., 1978, 43(18): 3481~3489

[26] Iumach G, Kiihle E. A new pathway to unsymmetrical disulfides. The thio-induced fragmentation of sulfenyl thiocarbonates. J. Am. Chem. Soc., 1970, 92(26): 7629~7631

[27] Swan J M. Thiols, disulphides and thiosulphates: some new reactions and possibilities in peptide and protein chemistry. Nature, 1957, (180): 634~644

[28] Hiskey R G. Synthesis of unsymmetrical aliphatic disulfides. J. Org. Chem., 1961, (26): 1152~1154

[29] Oae S. New simple synthesis of unsymmetrical disulfides and thiolsulfonates. Chem. Lett., 1977, 893~896

[30] Aiga Makoto, Samejima muneyasu. Quality improvement of organic isocyanate: 公开特许公报, 昭 59-172458. 1984. 9. 29

[31] Nakamura S, Nishimura O, Masuda H et al. Identification of volatile flavor components of roasted sesame oil//Bhattacharyya S C, Sen N, Sethi K L. 11th International Congress of Essential Oils, Fragrances and Flavours. Proceedings: Volume5, Chemistry—Analysis, Structure and Synthesis. New Delhi: Oxford & Ibh publishing Co. Pvt. Ltd., 1989. 73~87

[32] Werkhoff P, Brüning J, Emberger R et al. Flavor chemistry of meat volatiles: new results on flavor components from beef, pork, and chicken//Hopp R, Mori K. Recent Developments in Flavor and Fragrance Chemistry. New York: VCH Publisher, 1992. 183~213

[33] Mosciano G et al. Organoleptic characteristics of flavor materials. Perfumer & Flavorist, 1998, 23(5): 49~52

[34] Mosciano G et al. Organoleptic characteristics of flavor materials. Perfumer & Flavorist, 1993, 18(5): 39~41

[35] May C G. Thermal process flavorings//Ashurst P R. Food Flavorings. London: Blackie Academic & Professional, 1995. 276~314

[36] Oser B L, Hall R L. Recent progress in the consideration of flavoring ingredients under the food additives amendment, 5. GRAS Substances. Food Technology, 1972, 26(5): 36~41

[37] 孙宝国. 一种二糠基二硫醚的制备方法: 中国, ZL941005887. 1995. 1. 25

[38] Mottram D S. Meat//Maarse H. Volatile Compounds in Foods and Beverages. Zeist, The Netherlands: TNO-CIVO Food Analysis Institute, 1991. 137~139

[39] Piggott J R, Paterson A. Understanding Natural Flavor. New York: Blackie Academic & Professional, 1994. 157~159

[40] Teranishi R, Flath R A, Sugisawa H. Flavor Research. Recent Advances. New York: Marcel Dekker Inc., 1981. 222~224

[41] Hartman G J, Carlin J T, Scheide J D et al. Volatile products formed from the thermal degradation of thiamin at high and low moisture levels. J. Agric. Food Chem., 1984, 32: 1015~1018

[42] 郭新颜,孙宝国,宋焕禄. 在 Maillard 反应的肉味香精中二(2-甲基-3-呋喃基)二硫醚的鉴别. 化学通报, 2001, (1): 50~53

[43] Bedoukian P Z. Bedoukian's 47th annual reviem: Perfumery and flavor materials. Perfumer & Flavorist, 1991, 16(3): 1~27

[44] Evers W J et al. Furans substituted at the three position with sulfur//Charalambous G, Katz I. Phenolic, Sulfur, and Nitrogen Compounds in Food Flavors. Washington DC: American Chemical Society, 1976. 184~193

[45] 孙宝国,何坚,梁梦兰等. 一种二(2-甲基-3-呋喃基)二硫醚的制备方法: 中国, ZL941005879. 1995.1.25

[46] Oser B L, Ford R A. Recent progress in the consideration of flavoring ingredients under the food additives amendment, 11. GRAS Substances. Food Technology, 1978, 32(2): 60~70

[47] Smith R L, Cohen J, Doull J et al. GRAS flavoring substances 22. Food Technology, 2005, 59(8): 24~62

[48] Hall R L, Oser B L. Recent progress in the consideration of flavoring ingredients under the food additives amendment. 4. GRAS Substances. Food Technology, 1970, 24(5): 25~34

[49] Hall R L, Oser B L. Recent progress in the consideration of flavoring ingredients under the food additives amendment. 3. GRAS Substances. Food Technology, 1965, 19(2): 151~197

[50] Mosciano G, Sadural S, Fasano M et al. Organoleptic characteristics of flavor materials. Perfumer & Flavorist, 1989, 14(6): 47~55

[51] Oser B L, Ford R A. Recent progress in the consideration of flavoring ingredients under the food additives amendment. 12. GRAS Substances. Food Technology, 1979, 37(7): 65~73

[52] Mosciano G, Long J, Holmgten C et al. Organoleptic characteristics of flavor materials. Perfumer & Flavorist, 1996, 21(2): 47~49

[53] Güntert M, Bertram H J, Emberger R et al. Thermal degradation of thiamin(Vitamin B_1)//Mussinan C J, Keelan M E. Sulfur Compounds in Foods. Washington DC: American Chemical Society, 1994: 220~222

[54] Evers W J et al. 3-Furyl alkyl disulfides and foodstuff flavor composition comprising same: GB, 1538073. Jan. 17, 1979

[55] Evers W J, Heinsohn J, Howard H et al. 3-Furylalkyldisulfide, verfahren zu deren herstellung und ihre verwendung als wuerzstoffe: DE, 2604340A1. 1976.9.9

[56] 孙宝国,何坚,梁梦兰等. 一种烷基 2-甲基-3-呋喃基二硫醚的制备方法: 中国,

971006474. 1998.9.16

[57] 孙宝国,刘玉平,梁梦兰等. 一种甲基烃基二硫醚的制备方法:中国,ZL98100849.6. 1998.7.29

[58] 孙宝国等. 一种甲基2-甲基-3-呋喃基二硫醚的制备方法:中国,011000856. 2001.8.15

[59] 孙宝国等. 一种丙基2-甲基-3-呋喃基二硫醚的制备方法:中国,011000848. 2001.8.15

[60] Mosciano G et al. Organoleptic characteristics of flavor materials. Perfumer & Flavorist, 1997, 22(1): 57~59

[61] Morton I D. Food Flavors, Part A. Introduction. New York: Elsevier Science Publishing Company Inc., 1982. 237~239

[62] Leffingwell J C, Leffingwell D. GRAS Flavor chemicals—detection thresholds. Perfumer & Flavorist, 1991, 16(1): 1~19

[63] Oser B L, Ford R A. Recent progress in the consideration of flavoring ingredients under the food additives amendment. 6. GRAS Substances. Food Technology, 1973, 27(1): 64~68

[64] Milligan B, Swan J M. Unsymmetrical dialkyl disulfides from Bunte salts. J. Chem. Soc., 1963, (12): 6008~6012

[65] Smith R L, Doull J, Feron V J et al. GRAS flavoring substances 20. Food Technology, 2001, 55(12): 34~55

[66] Mosciano G. Organoleptic characteristics of flavor materials. Perfumer & Flavorist, 1996, 21(5): 49~54

[67] Oser B L, Ford R A. Recent progress in the consideration of flavoring ingredients under the food additives amendment. 10. GRAS Substances. Food Technology, 1977, 31(1): 65~74

[68] Wallace T J. Reaction of thiols with sulfoxides. I. Scope of the reaction and synthetic applications. J. Am. Chem. Soc., 1964, 86: 2018~2021

[69] Werkhoff P, Brüning J, Emberger R et al. Flavor chemistry of meat volatiles: new results on flavor components from beef, pork, and chicken//Hopp R, Mori K. Recent Developments in Flavor and Fragrance Chemistry. New York: VCH Publisher, 1992. 183~213

[70] Werkhoff P, Brüning J, Emberger R et al. Studies on volatile sulphur-containing flavour components in yeast extract//Bhattacharyya S C, Sen N, Sethi K L. 11th International Congress of Essential Oils, Fragrances and Flavours. Proceedings: Volume4, Chemistry Analysis and Structure. New Delhi: Oxford & Ibh publishing Co. Pvt. Ltd., 1989. 215~243

[71] Macleod G. The flavor of beef//Shahidi F. Flavor of Meat and Meat products. New York: Blackie Academic & Professional, 1994. 4~33

[72] Macleod G. The flavor of beef//Shahidi F. Flavor of Meat, Meat Products and Seafoods. London: Blackie Academic & Professional, 1998. 28~60

[73] Shaikh Y. Aroma chemicals in meat flavors. Perfumer & Flavorist, 1984, 9(3): 49~52

第六章 多硫醚类香料

多硫醚化合物的结构可以用下面的通式表示：

$$R-(S)_n-R'$$

式中：R、R'为烃基；$n \geq 3$。

多硫醚化合物广泛存在于各种食品中，如二甲基三硫存在于扇贝肉中[1]；二烯基三硫存在于大蒜油中[2]。目前香料工业中使用的多硫醚一般是三硫醚和四硫醚，有时允许使用几种多硫醚的混合物，如二烯丙基多硫醚（FEMA 号 3533）[3]。

第一节　多硫醚的一般合成方法

一、硫醇与二氯化一硫反应

硫醇与二氯化一硫反应是制取对称三硫醚的常用方法。例如[4]：

$$RSH + SCl_2 \longrightarrow RSSSR + HCl$$

二、硫醇与元素硫反应

硫醇与元素硫在胺存在下反应可以制备对称三硫醚。

$$RSH + S \xrightarrow{\text{胺}} RSSSR$$

三、烃硫基硫代碳酸酯与叔丁醇钾反应

烃硫基硫代碳酸酯与叔丁醇钾或叔丁醇钠反应可以制备对称三硫醚[5]。

$$\underset{O}{\overset{\parallel}{RSSCOCH_3}} + (CH_3)_3COK \longrightarrow RSSSR + \underset{O}{\overset{\parallel}{(CH_3)_3COCOCH_3}} + \underset{O}{\overset{\parallel}{KSCOCH_3}}$$

四、二咪唑基硫化物与硫醇反应

二咪唑基硫化物与硫醇反应可以制备对称三硫醚[6]。

$$\text{(Im-S-Im)} + RSH \longrightarrow RSSSR + \text{ImH}$$

五、硫代磺酸酯或硫代亚磺酸酯与二(三甲基硅)硫醚反应

硫代磺酸酯或硫代亚磺酸酯与二(三甲基硅)硫醚反应可以制备对称三硫醚[7,8]。例如：

$$\underset{\text{O}}{\overset{\text{O}}{\|}}RSSR + (CH_3)_3SiSSi(CH_3)_3 \longrightarrow RSSSR + (CH_3)_3SiOSi(CH_3)_3$$

$$\underset{\text{O}}{\overset{\text{O}}{\|}}RSSR\underset{\text{O}}{\overset{\text{O}}{\|}} + (CH_3)_3SiSSi(CH_3)_3 \longrightarrow RSSSR + R\overset{\text{O}}{\underset{\text{O}}{\|}}SOSi(CH_3)_3$$

六、Bunte 盐与硫化钠反应

Bunte 盐与硫化钠反应可以制备对称三硫醚[9,10]。例如：

$$RS_2O_3Na + Na_2S \longrightarrow RSSSR + Na_2SO_3$$

七、硫代磺酸酯与硫化钠或硫化钾反应

硫代磺酸酯与硫化钠或硫化钾反应可以制备对称三硫醚[11]。例如：

$$RSO_2SR_1 + K_2S \longrightarrow R_1SSSR_1 + RSO_2K$$

八、硫醇与二氯化二硫反应

硫醇与二氯化二硫反应是制取对称四硫醚的常用方法。例如[12~15]：

$$\text{iPr-SH} + S_2Cl_2 \longrightarrow \text{iPr-SSSS-iPr} + HCl$$

九、N-甲酯基腙与硫化氢反应

N-甲酯基腙与硫化氢反应可制备对称四硫醚[16]。

$$\underset{R_2}{\overset{R_1}{>}}C{=}O + NH_2NHCOOCH_3 \longrightarrow \underset{R_2}{\overset{R_1}{>}}C{=}NNHCOOCH_3$$

$$\xrightarrow[\text{CH}_3\text{OH}]{\text{H}_2\text{S}} \begin{array}{c} R_1 \\ | \\ \text{CSSSSC} \\ | \\ R_2 \end{array} \begin{array}{c} R_1 \\ | \\ \\ | \\ R_2 \end{array}$$

十、卤代烷与硫化氢和硫反应

卤代烷与硫化氢和硫反应，当硫化氢和硫的物质的量比为1:3时，主要生成对称四硫醚[17]。

$$RX + H_2S + S \xrightarrow{NH_3} RSSSSR + NH_4X$$

上述反应一般控制在碱性条件下进行。

十一、叔硫醇与硫反应

叔硫醇与硫反应可以制备四硫醚[18,19]。

$$\begin{array}{c} R_1 \\ | \\ R_2-C-SH \\ | \\ R_3 \end{array} + S \longrightarrow \begin{array}{c} R_1 \\ | \\ R_2-C \\ | \\ R_3 \end{array} -SSSS- \begin{array}{c} R_1 \\ | \\ C-R_2 \\ | \\ R_3 \end{array}$$

十二、二烷氧基二硫化物与硫醇反应

二烷氧基二硫化物与硫醇反应可以制备对称四硫醚或不对称四硫醚[20]。

$$ROSSOR + R_1SH \longrightarrow ROSSSR_1 \xrightarrow{R_2SH} R_2SSSSR_1$$

第二节 三硫醚类香料

一、二甲基三硫醚

二甲基三硫醚(dimethyl trisulfide，FEMA号3275)是白葡萄酒、威士忌酒、红茶、西红柿、卷心菜、花菜、花椰菜、洋葱、牛肉、海星、虾、鲐鱼[21~23]等的香成分。在玫瑰油中也发现了二甲基三硫醚[24]。该香料为无色至浅黄色液体，具有洋葱、大蒜、蔬菜、菜肴、肉香、鸡蛋香气以及葱、蒜、青香、薄荷、热带水果味道，可以用于洋葱、大蒜、大葱、细香葱、留兰香、菠萝、芒果、牛肉、蔬菜等食品香精。

二甲基三硫醚可以由硫酸二甲酯和硫代硫酸钠反应生成的Bunte盐与硫化钠反应制备[25]。

$$(CH_3)_2SO_4 + Na_2S_2O_3 \longrightarrow CH_3S_2O_3Na \xrightarrow{Na_2S} CH_3SSSCH_3$$

二、其他三硫醚香料

其他一些三硫醚香料的化学结构和香味特征见表6-1。

表6-1 其他一些三硫醚香料的化学结构和香味特征

名称	化学结构式	香味特征或应用建议
甲基乙基三硫醚（FEMA号3861）	CH₃-SSS-C₂H₅	可用于焙烤食品、软饮料、油脂、糖果、肉制品、奶制品、调味品等香精[26]
二乙基三硫醚（FEMA号4029）	C₂H₅-SSS-C₂H₅	洋葱香、大蒜香气[27]。可用于焙烤食品、饮料、糖果、肉制品、奶制品等香精[28]
二异丙基三硫醚（FEMA号3968）	iPr-SSS-iPr	可用于焙烤食品、饮料、油脂、糖果、肉制品、奶制品、调味品等香精[29]
甲基丙基三硫醚（FEMA号3308）	CH₃-SSS-C₃H₇	硫磺样、青香、大葱香、大蒜香[27]
乙基丙基三硫醚（FEMA号4042）	C₂H₅-SSS-C₃H₇	可用于焙烤食品、饮料、糖果、肉制品、奶制品等香精[28]
二丙基三硫醚（FEMA号3276）	C₃H₇-SSS-C₃H₇	洋葱香、大蒜香[27]
甲基烯丙基三硫醚（FEMA号3253）	CH₂=CHCH₂-SSS-CH₃	大蒜、洋葱香气[27]。可用于焙烤食品、肉制品、调味品、泡菜等香精[30]
二烯丙基三硫醚（FEMA号3265）	CH₂=CHCH₂-SSS-CH₂CH=CH₂	洋葱香、大蒜香[27]
4,5,6-三硫杂环庚烯	（环状S-S-S结构）	烤肉香
二（2-甲基-3-呋喃基）三硫醚	（2-甲基-3-呋喃基）-SSS-（2-甲基-3-呋喃基）	肉汤香[21]

名称	化学结构式	香味特征或应用建议
甲基 2-甲基-3-呋喃基三硫醚	(2-甲基-3-呋喃基)-SSS-甲基	肉香[31]

第三节 四硫醚类香料

一、二(2-甲基-3-呋喃基)四硫醚

二(2-甲基-3-呋喃基)四硫醚[bis(2-methyl-3-furyl)tetrasufide,FEMA 号 3260]为浅黄色液体,具有炖牛肉香气和味道[32],可用于调配牛肉、猪肉、鸡肉等肉味香精。

二(2-甲基-3-呋喃基)四硫醚可以通过 2-甲基-3-呋喃硫醇与二氯化二硫反应制备[33]。

$$\text{(2-methyl-3-furyl)-SH} + S_2Cl_2 \longrightarrow \text{(2-methyl-3-furyl)-SSSS-(2-methyl-3-furyl)} + HCl$$

二、其他四硫醚香料

其他一些四硫醚香料的化学结构和香味特征见表 6-2。

表 6-2 其他一些四硫醚香料的化学结构和香味特征

名称	化学结构式	香味特征
二甲基四硫醚	CH₃-SSSS-CH₃	猪肉挥发性香成分[34]
二丙基四硫醚	C₃H₇-SSSS-C₃H₇	硫化物样、洋葱香、大蒜香[35]
二异丙基四硫醚	(CH₃)₂CH-SSSS-CH(CH₃)₂	
二(2-甲基-3-呋喃基)四硫醚(FEMA 号 3260)	二(2-甲基-3-呋喃基)-SSSS-(2-甲基-3-呋喃基)	炖牛肉香[26]

参 考 文 献

[1] Flament I. Molecular gastronomy. Perfumer & Flavorist, 1997, 22(1): 1~8
[2] Rowe D. More fizz for your buck: High-impact aroma chemicals. Perfumer & Flavorist, 2000, 25(5): 1~19
[3] Oser B L, Ford R A. Recent progress in the consideration of flavoring ingredients under the food additives amendment. 11. GRAS Substances. Food Technology, 1978, 32(2): 60~70
[4] Decker Q W, Post H W. Preparation and ultraviolet absorption spectra of certain alkyl polysulfides. J. Org. Chem., 1957, 22: 145~146
[5] Harpp D. N, Granata A. Organic sulfur chemistry. XXI. Trisulfide formation by alkoxide decomposition of sulfenylthiocarbonates. Tetrahedron Lett., 1976, (35): 3001~3004
[6] Banerji A, Kalena G P. A new synthesis of organic trisulfides. Tetrahedron Lett., 1980, (31): 3003~3004
[7] Capozzi G. Capperucci A, Degl'Innocenti A et al. Silicon in organicsulphur chemistry. Part 1. Synthesis of trisulphides. Tetrahedron Lett., 1989, 30(22): 2991~2994
[8] Capozzi G, Capperucci A, Degl'Innocenti A et al. Synthesis of disulfides and trisulfides via organosilicon compounds. Gazz. Chim. Ital., 1990, 120(7): 421~426
[9] Milligan B, Swan J M. Unsymmetrical dialkyl disulfides from Bunts salts. J. Chem. Soc., 1963, (Dec.): 3608~3614
[10] Milligan B, Swan J M. Syntheses of thiooxamides and thioamides by exwension of the kindler reaction. J. Chem. Soc., 1961, 1194~2000
[11] Buckman J D, Field L. Organic disufides and related substances. XX. A novel preparation of symmetrical trisulfides using thiosulfonates. J. Org. Chem., 1967, 32(2): 454~457
[12] 刘玉平,孙宝国. 二异丙基四硫的合成研究. 精细化工,1998,15(1):28~30
[13] Chakravarti G C. Action of sulfur monochloride on mercaptans. J. Chem. Soc., 1923, 123: 964~968
[14] Clayton J O, Etzler D H. Hexadecyl trisulfide and hexadecyl tetrasulfide. J. Am. Chem. Soc., 1947, 69: 974~975
[15] Evers W J et al. Furans substituted at the three position with sulfur//Charalambous G, Katz I. Phenolic, Sulfur, and Nitrogen Compounds in Food Flavors. Washington DC: American Chemical Society, 1976. 184~193
[16] Ballini R. Reaction of ethoxycarbonylhydrazones with hydrogen sulfide: A new and facile synthesis of tetrasulfides. Synthesis, 1982, 9(10): 834~836
[17] Sperier J L. preparation of alkyl polysulfides: US, 4125552. 1978.11.14
[18] Bernard B, Edeard D. Process for the manufacture of dialkyl disufides and polyslfides: US, 4937385. 1990.6.26

[19] Nubar O. Production of dihydrocarbyl polysufides: US, 5146000. 1992.9.8

[20] Kagami H, Motoki S. Nucleophilic substitution on dialkoxy disulfide. Reactions with mercaptans or amines. J. Org. Chem., 1977, 42(25): 4139~4141

[21] Teranishi R, Flath R A, Sugisawa H. Flavor Research. Recent Advances. New York: Marcel Dekker Inc., 1981. 221~224

[22] Shahidi F, Cadwallader K R. Flavor and Lipid Chemistry of Seafoods. Washington DC: American Chemical Society, 1997. 47~66

[23] Belitz H D. Food Chemistry. Second Edition. New York: Springer, 1999. 336~337

[24] Boelens M H, Gemert L J V. Volatile character-impact sulfur compounds and their sensory properties. Perfumer & Flavorist, 1993, 18(3): 29~39

[25] 刘玉平,孙宝国,田红玉等. 二甲基三硫的合成研究//第三次全国精细化工青年科技学术交流会委员会. 第三次全国精细化工青年科技学术交流会论文集. 大连:大连出版社, 1998. 122~123

[26] Newberne P, Smith R L, Doull J et al. GRAS Flavoring substances 18. the 18th publication by the Flavor and Extract Manufactures' Association's expert panel on recent progress in the consideration of flavoring ingredients generally recognized as safe under the food additives amendment. Food Technology, 1998, 52(9): 79~92

[27] 孙宝国,刘玉平. 食用香料手册. 北京:中国石化出版社, 2004. 412~414

[28] Smith R L, Cohen S M, Doull J et al. GRAS Flavoring substances 21. Food technology, 2003, 57(5): 46~59

[29] Smith R L, Doull J, Feron V J et al. GRAS Flavoring substances 20. Food technology, 2001, 55(12): 34~55

[30] Oser B L, Hall R L. Recent progress in the consideration of flavoring ingredients under the food additives amendment, 5. GRAS Substances. Food Technology, 1972, 26(5): 36~41

[31] Güntert M, Bertram H J, Emberger R et al. Thermal degradation of thiamin(Vitamin B_1)//Mussinan C J, Keelan M E. Sulfur Compounds in Foods. Washington DC: American Chemical Society, 1994. 220~222

[32] Teranishi R, Flath R A, Sugisawa H, Flavor Research. Recent Advances. New York: Marcel Dekker Inc., 1981. 222~223

[33] Evers W J. Certain 3-furyl sulfides: US, 4020175. 1977.4.26

[34] Werkhoff P, Brüning J, Emberger R et al. Flavor chemistry of meat volatiles: new results on flavor components from beef, pork, and chicken//Hopp R, Mori K. Recent Developments in Flavor and Fragrance Chemistry. New York: VCH Publisher, 1992. 201~203

[35] 刘玉平,孙宝国,田红玉等. 二正丙基四硫合成的研究. 北京轻工业学院学报, 1998, 16(3):6~8

第七章 一硫代缩羰基类香料

第一节 一硫代缩醛的一般合成方法

一硫代缩醛(O,S-缩醛)是缩醛的一个烷氧基被烷硫基取代的产物,结构特点是在同一个碳原子上连有一个烷氧基和一个烷硫基。

$$\begin{array}{c} R' \diagdown \diagup OR_1 \\ C \\ H \diagup \diagdown SR_2 \end{array}$$

一硫代缩醛(O,S-缩醛)

由于一硫代缩醛其结构上的特殊性,导致其性能上的特殊性,在有机合成中已成为一类非常有用的化合物[1~3],一硫代缩醛作为羰基保护基团[4,5]和反应中间体的重要性已经引起了人们的关注[6~10]。

一硫代缩醛化合物在有机合成上具有许多应用[11],如烷基立体选择合成[12]、四氢呋喃的立体控制合成[13]、1,3-二醇的立体合成[14]、[2.3]维梯希加成[13,15,16]、反式布鲁克重排[17]等。

在香料工业领域,一硫代缩醛类中的 2-甲基-4-丙基-1,3-氧硫杂环己烷(FEMA号 3578)已被允许作为食品香料使用,在焙烤食品、奶制品、肉制品、软糖中的用量一般为 0.1mg/kg,在饮料中的用量一般为 0.05mg/kg[18]。此类化合物中的 1,3-氧硫杂环戊烷类化合物分子结构中含有肉香味含硫化合物特征结构单元,一般具有肉类香味[19],是一类有发展前景的肉味香料。

以下介绍一硫代缩醛的主要合成方法。

一、硫醇与烯醚加成

烯醚与硫醇或硫酚加成可以选择性地直接合成 O,S-缩醛[20,21]。常用的溶剂是四氢呋喃、二氯甲烷,催化剂有 $AlCl_3$、$ZnCl_2$、$HgCl_2$、$TiCl_4$ 等。

$$\underset{R'}{\overset{OR}{\diagup}}\!\!=\!\!R'' + R'''SH \xrightarrow[H_3O^+]{THF} R'\overset{OR}{\underset{SR'''}{|}}R''$$

用烷基硫醇反应时产率一般较低,用苯硫酚和糠硫醇等亲核性较强的含硫试剂反应时产率较高。

二、二甲亚砜与醇反应

二甲亚砜与醇反应主要用于制备 1-烷氧基-1-烷硫基甲烷化合物,如正丁氧基甲硫基甲烷[22,23]。

$$(CH_3)_2SO + BuOH \xrightarrow[2) NaOH/(C_2H_5)_2O]{1) (CH_3CO)_2O} BuO\diagup S\diagdown + H_2O$$

三、由醛合成

1. 醛与三甲基硅氧烷及 Me₃SiSPh 反应

在 $Me_3SiOSO_2CF_3$(TMSOTf)存在下,醛与三甲基硅氧烷反应可以合成 O,O-缩醛[24]。在反应中调节 TMSOTf 的用量,以 CH_2Cl_2 作溶剂,醛与相同比例的 Me_3SiSPh 和三甲基硅氧烷在 $-78℃$ 下反应可以得到以 O,S-缩醛为主的反应产物。

$$\underset{R'}{\overset{O}{\|}}\underset{H}{C} + Me_3SiOR'' + Me_3SiSPh \xrightarrow[CH_2Cl_2]{TMSOTf} \underset{R'}{\overset{OR''}{\underset{H}{C}}}\underset{}{SPh}$$

2. 醛和硫醇(硫酚)、醇反应

醛与硫醇(硫酚)和醇反应,可以制备 O,S-缩醛。例如[25]:

$$PhCHO + PhSH + C_2H_5OH \longrightarrow Ph\underset{SPh}{\overset{OEt}{CH}}$$

此方法用亲核性较强的苯硫酚和糠硫醇效果较好。

四、由缩醛合成

1. 由 α,β-不饱和缩醛合成

α,β-不饱和缩醛,如(E)-1,1-二甲氧基-2-丁烯,以四氢呋喃为溶剂,在二甲基硫醚和 $Me_3SiOSO_2CF_3$(TMSOTf)存在下,在 $-78℃$ 下反应一定时间,然后与苯硫酚锂反应,得到反式 1-苯硫基-1-甲氧基-2-丁烯。

通过硅胶气相色谱分离可以得到产率为 92.0% 的产品[26]。该方法可以合成

含有不饱和双键的 O,S-缩醛。

2. 缩醛与硫醇在路易斯酸催化下反应

缩醛与硫醇(硫酚)以路易斯酸作催化剂进行反应,缩醛的一个烷氧基被烷(苯)硫基取代[27]。

$$\underset{R''}{\overset{R'}{>}}C\underset{OMe}{\overset{OMe}{<}} + PhSH \xrightarrow{\text{路易斯酸}} \underset{R''}{\overset{R'}{>}}C\underset{SPh}{\overset{OMe}{<}} + \underset{R''}{\overset{R'}{>}}C\underset{SPh}{\overset{SPh}{<}}$$

3. 缩醛与有机锡硫化物反应

缩醛与有机锡硫化物在甲苯和三氟化硼·乙醚溶液中反应,用气相色谱分析反应产物。结果表明,90%以上的为一硫代缩醛,没有发现缩硫醛[27,28]。常用的有机锡硫化物为 $Bu_nSn(SR)_n$($n=1$ 或 2;R=Ph 或 $PhCH_2$)。

$$\underset{R''}{\overset{R'}{>}}C\underset{OR'''}{\overset{OR'''}{<}} + Bu_nSn(SR)_n \xrightarrow[\text{—CH}_3]{BF_3·OEt_2} \underset{R''}{\overset{R'}{>}}C\underset{SR}{\overset{OR'''}{<}}$$

4. 缩醛与有机铝硫化物反应

有机铝硫化物 R_2AlSR' 作为一种有效的亲核试剂,能使 C—O 键断裂,用 SR 亲核基团取代,生成一硫代缩醛[29]。例如:

$$\underset{R'''}{\overset{R''}{>}}C\underset{OMe}{\overset{OMe}{<}} + Et_2AlSPh \longrightarrow \underset{R'''}{\overset{R''}{>}}C\underset{SPh}{\overset{OMe}{<}}$$

5. 缩醛与硫醇在溴化镁催化下反应

缩醛与硫醇以乙醚作溶剂,无水溴化镁作催化剂,在常温下反应,产物色谱分析表明得到产率很好的一硫代缩醛[30]。

$$\underset{R''}{\overset{R'}{>}}C\underset{OR'''}{\overset{OR'''}{<}} + RSH \xrightarrow[Et_2O]{MgBr_2} \underset{R''}{\overset{R'}{>}}C\underset{SR}{\overset{OR'''}{<}}$$

此方法用亲核性强的苯硫酚和糠硫醇效果较好。

6. 2-烷氧基四氢吡喃与硫醇反应

在二甲基溴化硼催化下,2-烷氧基四氢吡喃与硫醇反应得到高产率的 O,S-缩醛[31,32]。

$$\text{（四氢吡喃-OR）} + \text{EtSH} \xrightarrow[\text{CH}_2\text{Cl}_2]{\text{Me}_2\text{BBr}} \text{HO—...—EtS—OR}$$

常用的溶剂是二氯甲烷、四氢呋喃等。

7. 缩醛和硫醇在二氰基乙烯酮缩酮催化下反应

在一种新型 π-酸催化剂——二氰基乙烯酮缩酮[如二氰基乙烯酮二甲缩酮（DCKMA）和二氰基乙烯酮二乙烯缩酮（DCKEA）]催化下，α,β-不饱和缩醛与硫醇在温和反应条件下反应生成 O,S-缩醛[33,34]。这种催化剂可以由四氰基乙烯制备，是区别于路易斯酸的一种新型 π-酸催化剂，它对于 α,β-不饱和缩醛化合物直接转化为相应的 O,S-缩醛化合物尤为有效。

$$\underset{R''}{\overset{R'}{>}}C\underset{\text{OMe}}{\overset{\text{OMe}}{<}} + \text{PhSH} \xrightarrow[\text{DMF}]{\text{DCKEA}} \underset{R''}{\overset{R'}{>}}C\underset{\text{SPh}}{\overset{\text{OMe}}{<}}$$

8. 缩醛和硫醇在四氯化碳中反应

缩醛和硫醇在四氯化碳中反应可以合成许多一硫代缩醛类化合物[35~37]。

$$\underset{R''}{\overset{R'}{>}}C\underset{\text{OR}'''}{\overset{\text{OR}'''}{<}} + \text{RSH} \xrightarrow{\text{CCl}_4} \underset{R''}{\overset{R'}{>}}C\underset{\text{SR}}{\overset{\text{OR}'''}{<}}$$

五、由 α-卤代醚合成

α-卤代醚与硫醇在二甲基甲酰胺（DMF）中反应可以得到一硫代缩醛化合物[38]。

$$\text{Ph—CH(OMe)(Cl)} + \text{X—C}_6\text{H}_4\text{—SH} \xrightarrow{\text{DMF}} \text{Ph—CH(OMe)—S—C}_6\text{H}_4\text{—X}$$

六、由环状缩硫醛合成

环状缩硫醛先与氟磺酸甲酯反应得到硫鎓离子，然后再与巯基乙醇或巯基丙醇反应得到环状一硫代缩醛[39]，产率一般高于 70%。例如：

$$\underset{H}{\overset{R}{>}}C\underset{S}{\overset{S}{\diagup\diagdown}} + \text{CH}_3\text{SO}_3\text{F} \longrightarrow \underset{H}{\overset{R}{>}}C\underset{S}{\overset{\overset{+}{S}-\text{CH}_3}{\diagup\diagdown}} \text{FSO}_3^- \xrightarrow[\text{CH}_2\text{Cl}_2]{\text{HS} \atop \text{HO}} \underset{H}{\overset{R}{>}}C\underset{O}{\overset{S}{\diagup\diagdown}}$$

七、醛与巯基醇反应

醛与巯基醇(如巯基乙醇、巯基丙醇)反应可以合成环状一硫代缩醛[40]。

$$R-\underset{H}{\underset{|}{C}}=O + \begin{matrix}HS\\HO\end{matrix}\Big[\xrightarrow{CH_3-\text{\Large$\bigcirc$}-SO_3H} \underset{H}{\underset{|}{\overset{R}{\underset{|}{C}}}}\overset{S}{\underset{O}{\diagdown}}\Big] + H_2O$$

此方法操作简单,但受到巯基醇品种的限制。

第二节 开链一硫代缩醛类香料

开链一硫代缩醛可以用缩醛和硫醇为原料在四氯化碳溶液中反应合成。例如[37]:

$$\text{CH}_3\text{CH}(OC_2H_5)_2 + HS-C_3H_7 \xrightarrow{CCl_4} \text{CH}_3\text{CH}(OC_2H_5)(SC_3H_7) + HOC_2H_5$$

部分开链一硫代缩醛类香料的化学结构和沸点见表7-1。

表7-1 部分开链一硫代缩醛类香料的化学结构和沸点

名称	化学结构式	沸点
1-丙氧基-1-糠硫基甲烷	(结构式)	130~132℃/4.9kPa[41]
1-乙氧基-1-乙硫基乙烷	(结构式)	78~80℃
1-乙氧基-1-丙硫基乙烷	(结构式)	96~97℃/11.1kPa[37]
1-乙氧基-1-丁硫基乙烷	(结构式)	77~79℃/2.7kPa[37]
1-乙氧基-1-(1-甲基-1-丙硫基)乙烷	(结构式)	104~106℃/11.0kPa[37]
1-乙氧基-1-戊硫基乙烷	(结构式)	79~82℃/2.0kPa

续表

名称	化学结构式	沸点
1-乙氧基-1-糠硫基乙烷		86~89℃/1.2kPa[36,42]
1-乙氧基-1-苄硫基乙烷		136~140℃/1.1kPa
1-甲氧基-1-(1-甲基-1-丙硫基)丁烷		96~98℃/2.6kPa
1-甲氧基-1-丁硫基丁烷		114~116℃/1.5kPa
1-甲氧基-1-戊硫基丁烷		114~116℃/1.6kPa
1-甲氧基-1-糠硫基丁烷		140~141℃/3.8kPa
1-甲氧基-1-苄硫基丁烷		162~164℃/5.9kPa
1-乙氧基-1-丙硫基丁烷		64~66℃/1.3kPa
1-乙氧基-1-丁硫基丁烷		81~83℃/1.2kPa
1-乙氧基-1-戊硫基丁烷		127~128℃/2.9kPa
1-乙氧基-1-糠硫基丁烷		131~134℃/2.1kPa
1-乙氧基-1-苄硫基丁烷		130~131℃/0.1kPa

关于开链一硫代缩醛的合成孙宝国等也进行过一些研究,参见文献[43~46]。

第三节 环状一硫代缩醛类香料

2-甲基-4-丙基-1,3-氧硫杂环己烷（FEMA 号 3578）又称 3-巯基-1-己醇缩乙醛，是允许使用的环状一硫代缩醛香料。该化合物天然存在于西番莲中，具有热带水果香味，是热带水果香精的关键香料[47]，主要用于烤香、焦香、奶味、肉味、热带水果等食品香精，在最终加香食品中浓度为 0.05~0.1mg/kg。

2-甲基-4-丙基-1,3-氧硫杂环己烷可以通过下面的方法合成：

$$CH_3CH_2CH_2CHCH_2CH_2OH + CH_3CHO \longrightarrow \text{（结构式）} + H_2O$$
$$\quad\quad\quad\quad\quad |$$
$$\quad\quad\quad\quad SH$$

醛与 β-巯基醇反应生成的 1,3-氧硫杂环戊烷化合物大都具有肉香味，是一类很有希望新香料，部分此类化合物的化学结构和香气特征见表 7-2。

表 7-2 部分 1,3-氧硫杂环戊烷和环己烷化合物的化学结构和香气特征

名称	化学结构式	香气特征
1,3-氧硫杂环戊烷	（结构式）	浓郁的葱蒜香、烤肉香气[48]
2-甲基-1,3-氧硫杂环戊烷	（结构式）	烤肉香、浓郁的葱蒜香气[49]
2-乙基-1,3-氧硫杂环戊烷	（结构式）	葱蒜香香气，烤香气[50]
2-丙烯基-1,3-氧硫杂环戊烷	（结构式）	浓郁的烤肉香和青叶香气[50]
2-正丙基-1,3-氧硫杂环戊烷	（结构式）	烤肉香、葱蒜香、辛香香气[49]
2-正丁基-1,3-氧硫杂环戊烷	（结构式）	烤肉香、辛香和葱蒜香香气[48]
2-(2-呋喃基)-1,3-氧硫杂环戊烷	（结构式）	浓郁的炖肉香及咖啡香气[51]

续表

名称	化学结构式	香气特征
2-苯基-1,3-氧硫杂环戊烷	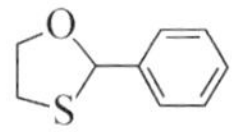	浓郁的烤肉香及苦杏仁香[51]
2-正己基-1,3-氧硫杂环戊烷	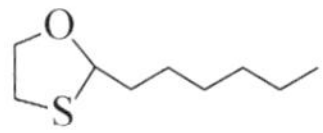	烤肉香、辛香、葱蒜香气[48]
2-正庚基-1,3-氧硫杂环戊烷	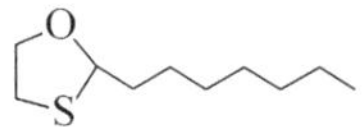	烤肉香、辛香和葱蒜香气[48]
4-(1,3-氧硫杂-2-环戊烷基)正丁醛	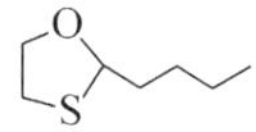	浓郁的青叶香,烤肉香[48]
1,3-二(1,3-氧硫杂-2-环戊烷基)丙烷	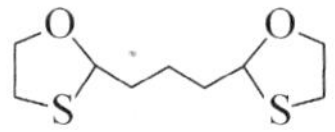	烤肉香气,弱青香气[48]
4,4-二甲基-2-乙基-1,3-氧硫杂环己烷	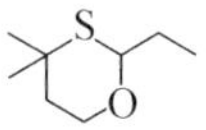	药草、鼠尾草、万寿菊、柑橘、野薄荷香气;硫磺气息,洋葱后韵[52]
4,4-二甲基-2-庚基-1,3-氧硫杂环己烷	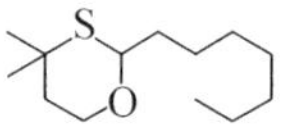	果香、黑醋栗芽香、柠檬香气[52]
4,4-二甲基-2-(2,4-二甲基-3-环己烯-1-基)-1,3-氧硫杂环己烷	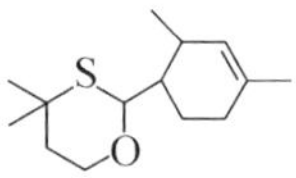	青香、药草香、山萝卜香、艾蒿香、万寿菊香、鼠尾草香气[52]

关于环状一硫代缩醛的合成孙宝国等也进行过一些研究,参见文献[48~51,53~55]

第四节 环状一硫代缩酮类香料

酮与巯基醇反应是合成环状一硫代缩酮化合物的有效方法。例如[48]:

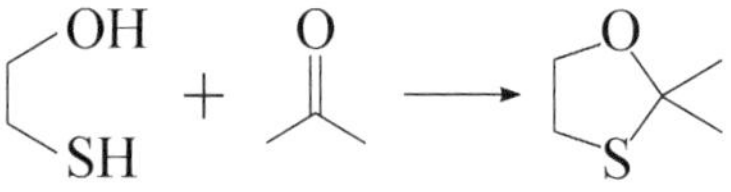

1,3-氧硫杂环戊烷类化合物一般具有葱蒜和肉类香味特征,关于这类化合物的研究已经引起了香料化学界的重视。表7-3列举了包括此类化合物在内的部分环状一硫代缩酮类香料的化学结构和香味特征。

表 7-3 部分环状一硫代缩酮类香料的化学结构和香味特征

名称	化学结构式	香味特征
2,2-二甲基-1,3-氧硫杂环戊烷		浓郁的葱蒜香、烤肉香气[48]
2-甲基-2-乙基-1,3-氧硫杂环戊烷		葱蒜香、烤肉香[48]
2-甲基-2-正丙基-1,3-氧硫杂环戊烷		烤香并略带葱蒜香[55]
2-甲基-2-正戊基-1,3-氧硫杂环戊烷		浓郁的草莓香和烤肉香气[48]
1-氧杂-4-硫杂螺[4,4]壬烷		蒜香、烤肉和青叶香气[48]
2-甲基-2-苯基-1,3-氧硫杂环戊烷		杏仁香、青叶和烤肉香[48]
2-乙基-2-苯基-1,3-氧硫杂环戊烷		杏仁香气、烤肉香气[48]
2-甲基-2-乙酰基-1,3-氧硫杂环戊烷		炖鸡肉香气、烤香气[48]
1-(2-甲基-1,3-氧硫杂-2-环戊烷基)丙酮		药香、烤肉香气[48]
1-(2-甲基-1,3-氧硫杂-2-环戊烷基)乙酸乙酯		苹果酯香、烤肉香、药香[48]
2,4,4-三甲基-2-戊基-1,3-氧硫杂环己烷		甜的、蜜香香气,稍有洋葱香气,罗勒叶、鼠尾草、万寿菊香气,硫化物气息[52]

参 考 文 献

[1] Olsen R K, Currie J O. The Chemistry of the Thiol Group, part 2. New York: Wiley. 1974. 520～523

[2] Kusche A, Hoffmann R, Münster I et al. A direct synthesis of O,S-acetals from aldehydes. Tetrahedron Lett., 1991, 32(4): 467~470

[3] Kim S, Park J H, Lee J M. Facile preparation of α,β-unsaturated O,S-acetals and mixed acetals via 3-alkoxy-2-alkenylenesulfonium salts. Tetrahedron Lett., 1993, 34(36): 5769~5772

[4] Greene T W. Protective Groups in Organic Synthesis. New York: John Wiley & Sons, 1981. 1~349

[5] Corey E J, Bock M G. Protection of primary hydroxyl groups as (methylthio) methyl ethers. Tetrahedron Lett., 1975, (38): 3269~3270

[6] Nicolaou K C, Caulfield T J, Kataoka H et al. Total synthesis of the tumor-associated Le^x family of glycosphingolipids. J. Am. Chem. Soc., 1990, 112: 3693~3695

[7] Veeneman G H, Boom J H V. An efficient thioglycoside-mediated formation of α-glycosidic linkages promoted by iodonium dicollidine perchlorate. Tetrahedron Lett., 1990, 31: 275~278

[8] Konradsson P, Udodong U E, Fraser-Reid B. Iodonium promoted reactions of disarmed thioglycosides. Tetrahedron Lett., 1990, 31: 4313~4316

[9] Jain R K, Matta K L. Methyl 3,4-O-isopropylidene-2-O-(4-methoxy-benzyl)-1-thio-β-L-fucopyranoside-a novel, efficient glycosylating reagent for the synthesis of linear and other α-L-fucosyl oligosaccharides. Tetrahedron Lett., 1990, 31: 4325~4328

[10] Kihlberg J O, Leigh D A, Bundle D R. The in situ activation of thioglycosides with bromine: An improved glycosylation method. J. Org. Chem., 1990, 55: 2860~2863

[11] Cohen T, Matz J R. A general preparative method for α-lithioethers and its application to a concise, practical synthesis of brevicomin. J. Am. Chem. Soc., 1980, 102: 6900~6902

[12] Cohen T, Bhupathy M. Organoalkali compounds by radical anion induced reductive metalation of phenyl thioethers. Acc. Chem. Res., 1989, 22: 152~161

[13] Broka C A, Shen T. Reductive lithiation mediated anionic cyclizations and [2,3]-sigmatropic rearrangements. J. Am. Chem. Soc., 1989, 111: 2981~2984

[14] Rychnovsky S D. A convergent synthesis of polyol chains. J. Org. Chem., 1989, 54: 4982~4984

[15] Kruse B, Brücker R. Reductive lithiation of sulfides and sulfones-a novel entry into [2,3] Wittig rearrangements. Chem. Ber., 1989, 122: 2023~2025

[16] Kruse B, Brücker R. Does the wittig-still rearrangement proceed via a metal free carbanion. Tetrahedron Lett., 1990, 31: 4425~4428

[17] Pedretti V, Veyrieres A, Sinay P. Synthesis of α-D-glycopyranosyl trimethylsilanes. Tetrahedron. 1990, 46: 77~88

[18] Oser B L, Ford R A. Recent progress in the consideration of flavoring ingredients under the food additives amendment. 11. GRAS Substances. Food Technology. 1978, 32(2): 60~70

[19] Sun B G, Miao C F, Yang Y Q et al. Aroma characteristics of 1,3-oxathiolanes// Yang J Z, Peng X J. The Proceedings of The 2nd International Conference on Functional Molecules. Dalian:Dalian University of Technology Press, 2003. 377~379

[20] Mukaiyama T, Izawa T, Saigo K et al. Addition reaction of thiol to olefin by the use of $TiCl_4$. Chemistry Letters, 1973, (4): 355~356

[21] Braga A L, Silveira C C, Dornelles L et al. Catalyst-dependent selective synthesis of O/S-acetals from enol ethers. Synthetic Communications, 1995, 25(20): 3155~3162

[22] Pojer P M, Angyal S J. Methylthiomethyl ethers: their use in the protection and methylation of hydroxyl groups. Australian J. Chem., 1978, 31: 1031~1040

[23] Jones J H, Thomas D W, Thomas R M et al. t-Butyl chloromethyl ether. Synthetic Communications, 1986, 16(13): 1607~1610

[24] Glass R S. Facile synthesis of trimethylsilyl thioethers. J. Organometal Chem., 1973, 61: 83~90

[25] Ferraz J P, Cordes E H. O-Ethyl S-phenyl benzaldehyde acetal hydrolysis. J. Am. Chem. Soc., 1979, 101(6): 1488~1491

[26] Kim S, Park J H, Lee J M. Facile preparation of α,β-unsaturated O,S-acetals and mixed acetals via 3-alkoxy-2-alkenylenesulfonium salts. Tetrahedron Lett., 1993, 34(36): 5769~5772

[27] Sato T, Otera J, Nozaki H. Activation and synthetic applications of thiostannanes. Thioalkoxylation of acetals. Tetrahedron, 1989, 45(4): 1209~1218

[28] Sato T, Kobayashi T, Gojo T et al. Organotin-mediated preparation of monothioacetals. Chem. Lett., 1987, (8): 1661~1664

[29] Masaki Y, Serizawa Y, Kaji K. Monothioacetalization of acetals using diethylaluminium thiophenoxide. Chem. Lett., 1985, (12):1933~1936

[30] Kim S, Park J H, Lee S. Conversion of acetals into monothioacetals, α-Alkoxyazides and α-alkoxyalkyl thioacetates with magnesium bromide. Tetrahedron. Lett., 1989, 30(48): 6697~6700

[31] Morton H E, Guindon Y. Dimethylbroron bromide interconversion of protecting groups: Preparation of MTM ethers, O,S-acetals, and cyanomethyl ethers. J. Org. Chem., 1985, 50: 5379~5382

[32] Guindon Y, Bernstein M A, Anderson P C. Ring cleavage of THP and THF ethers using dimethylboron bromide. Tetrahedron Lett., 1987, 28(20): 2225~2228

[33] Miura T, Masaki Y. Dicyanoketene acetals, a novel type of π-acid catalyst for monthioacetalization of acetals. Tetrahedron Lett., 1994, 35: 7961~7964

[34] Miura T, Masaki Y. Monothioacetalization of acetals catalyzed by dicyanoketene acetals. Tetrahedron, 1995, 51(38): 1047~10486

[35] 孙宝国,付翔,刘玉平等.一类1-烷基-1-烷氧基-1-烷硫基甲烷类化合物及其制备方法:中国,ZL001076361. 2000.1.17

[36] 付翔,孙宝国,郑福平等. 1-甲基-1-乙氧基-1-糠硫基甲烷合成研究.精细化工,2000, 17(12):717～718,732

[37] 付翔,孙宝国,郑福平等. 1-甲基-1-乙氧基-1-烷硫基甲烷类化合物的合成研究.化学通报,2001,(2):112～115

[38] Fife T H, Anderson E. Thioacetal hydrolysis. The hydrolysis of benzaldehyde methyl S-(substituted phenyl) thioacetals. J. Am. Chem. Soc., 1970, 92(18):5464～5466

[39] Corey E J, Hase T. Thioacetal-hemithioacetal-acetal interchange under mild conditions. Tetrahedron Lett., 1975, 38:3267～3268

[40] Carl D, Marvin G. Studies in organic sulfur compounds. Ⅵ. Cyclic ethylene and trimethylene hemithioketals. J. Am. Chem. Soc., 1953, 75:3704～3708

[41] 侯乐山,郑福平,孙宝国等. 1-丙氧基-1-糠硫基甲烷的合成工艺研究.精细化工,2001, 18(10):586～587,594

[42] 侯乐山,郑福平,孙宝国. 一种新的合成 1-乙氧基-1-糠硫基乙烷的方法//第四届全国青年精细化工学术会议委员会. 第四届全国青年精细化工学术会议论文集,天津:天津大学,2001. 138～141

[43] 杨迎庆,郑福平,孙宝国等. 1-乙氧基-1-仲丁硫基丙烷及丙醛二仲丁硫基缩硫醛的合成//第四届全国青年精细化工学术会议委员会. 第四届全国青年精细化工学术会议论文集,天津:天津大学,2001. 150～153

[44] 杨迎庆,郑福平,孙宝国等. 四种 1-乙氧基-1-烷硫基丙烷类化合物的合成. 化学试剂, 2002, 24(1):5～7,48

[45] 侯乐山,郑福平,孙宝国等. 1-甲氧基-1-烷硫基丙烷的合成.应用化学,2002,19(2): 179～180

[46] 侯乐山,郑福平,孙宝国等. 五种 1-烷基-1-烷氧基-1-烷硫基甲烷类化合物的合成研究. 化学通报(网络版),2001,(9):w01096

[47] Rowe D. More fizz for your buck: High-impact aroma chemicals. Perfumer & Flavorist, 2000, 25(5):1～19

[48] 孙宝国,杨迎庆,郑福平等. 15 种 1,3-氧硫杂环戊烷类香料化合物的合成.化学通报, 2002,65(9):614～619

[49] 杨迎庆,郑福平,孙宝国等. 2-甲基-1,3-氧硫杂环戊烷和 2-丙基-1,3-氧硫杂环戊烷的合成. 精细化工,2001,18(9):525～527

[50] Sun B G, Miao C F, Yang Y Q et al. Aroma characteristics of 1,3-oxathiolanes// Yang J Z, Peng X J. The Proceedings of The 2nd International Conference on Functional Molecules. Dalian:Dalian University of Technology Press, 2003. 377～379

[51] Yang Y Q, Zheng F P, Sun B G et al. Synthesis of three 2-alkyl(aryl)-1,3-oxathiolane compounds. Chemical Journal on Internet, 2001, 3(11)

[52] 舒宏福. 3-巯基脂肪醇和它的衍生物香料//中国香料香精化妆品工业协会. 2004 年中国香料香精学术研讨会论文集. 北京:中国香料香精化妆品工业协会,2004. 190～198

[53] 谢建春,孙宝国,郑福平等. 溴化镁催化合成 2-(2',6'-二甲基-1',5'-庚二烯基)-4,5-二甲

基-1,3-氧硫杂环戊烷. 化学试剂,2005,27(5):269~270,308
[54] Zheng F P, Sun B G, Liu Y P et al. Syntheses of six novel 1,3-oxathiolane flavoring compounds//Yang J Z, Peng X J. The Proceedings of The 2nd International Conference on Functional Molecules. Dalian: Dalian University of Technology Press,2003.284~286
[55] 杨迎庆,郑福平,孙宝国等. 2-甲基-2-丙基-1,3-氧硫杂环戊烷的合成. 化学试剂,2001,23(6):342~343

第八章 二硫代缩羰基类香料

二硫代缩羰基类化合物可以用下面的结构通式表示：

$$\mathrm{R_1\atop R_2}\!\!>\!\!\mathrm{C}\!\!<\!\!{SR_3\atop SR_4}$$

式中：R_1，R_2 为氢或烃基；R_3，R_4 为烃基。

在许多食品的挥发性香成分检测中发现了二硫代缩羰基化合物，如在煮牛肉中发现了1,3-二硫杂环戊烷、2-甲基-1,3-二硫杂环戊烷、1,3-二噻烷；在高压煮牛肉中发现了二甲硫醇缩乙醛；在奶酪、罐头牛肉中发现了二甲硫基甲烷[1]。二硫代缩羰基类化合物的化学性质比较稳定，香味特征性强，是一类很有发展前途的新型含硫香料。

第一节 二硫代缩羰基化合物的一般合成方法

一、醛(酮)与硫醇反应

醛(酮)与硫醇直接反应生成缩硫醛(酮)。

$$\mathrm{R_1\atop R_2}\!\!>\!\!\mathrm{C}\!\!=\!\!\mathrm{O} + R_3SH \xrightarrow{H^+} \mathrm{R_1\atop R_2}\!\!>\!\!\mathrm{C}\!\!<\!\!{SR_3\atop SR_3} + H_2O$$

常用的催化剂有 HCl[2]、HNO_3、H_2SO_4[3~16]、$FeCl_3$[17]、$AlCl_3$[17~19]、$TiCl_4$[17]、$TeCl_4$[17]、无水氯化锌[20,21]、$MgBr_2$[22]、$LiBr$[23]、$MgBr_2/Et_2O$[24]、蒙脱土 KSF 催化剂[25]、对甲苯磺酸[26~28]、$LiOTf$[29]、磺酸型离子交换树脂[30]、$PPSE$[31]、$SOCl_2$-SiO_2[32]、Me_3SiCl[3]等。

二、醛(酮)与二硫醚反应

醛(酮)与二硫醚反应可以制备缩硫醛(酮)[33]。

$$\mathrm{R_1\atop R_2}\!\!>\!\!\mathrm{C}\!\!=\!\!\mathrm{O} + R_3SSR_3 \xrightarrow{Bu_3P} \mathrm{R_1\atop R_2}\!\!>\!\!\mathrm{C}\!\!<\!\!{SR_3\atop SR_3} + Bu_3P\!\!=\!\!O$$

三、醛(酮)与 Bunte 盐反应

卤代烷和硫代硫酸钠反应生成的 Bunte 盐与醛(酮)反应可以制备缩硫醛(酮)[34,35]。

$$\text{CH}_3\text{CH}_2\text{CH}_2\text{CH}_2\text{Br} + \text{Na}_2\text{S}_2\text{O}_3 \longrightarrow \text{CH}_3\text{CH}_2\text{CH}_2\text{CH}_2\text{S}_2\text{O}_3\text{Na} \xrightarrow{\text{CH}_3\text{COC}_2\text{H}_5} \text{(RS)}_2\text{C(CH}_3\text{)(C}_2\text{H}_5\text{)}$$

四、醛(酮)与三烷硫基硼烷反应

醛(酮)与三烷硫基硼烷反应生成缩硫醛(酮),如糠醛与三甲硫基硼烷反应生成二甲硫缩糠醛[36]。

$$\text{furfural-CHO} + \text{B(SCH}_3\text{)}_3 \longrightarrow \text{furfural-CH(SCH}_3\text{)}_2$$

五、醛(酮)与有机硫磷化物反应

醛(酮)与有机硫磷化物反应可以制备 α,β-不饱和缩硫醛(酮)[37~40]。

$$\underset{R_1\ R_2}{\overset{O}{\|}}\text{C} + (\text{RO})_3\text{P=C(SR)}_2 \longrightarrow R_1R_2\text{C=C(SR)}_2$$

六、醛(酮)与三硫代原甲酸酯反应

原甲酸三乙酯与硫醇反应生成三硫代原甲酸三乙酯,再与醛(酮)缩合可以制备 α,β-不饱和缩硫醛(酮)[4,25,41]。

$$\text{HC(OC}_2\text{H}_5\text{)}_3 + \text{RSH} \longrightarrow \text{HC(SR)}_3 \xrightarrow{R_1R_2\text{C=O}} R_1R_2\text{C=C(SR)}_2$$

七、缩醛(酮)与硫醇反应

缩醛(酮)与硫醇发生交换反应生成缩硫醛(酮)。

$$R_1R_2\text{C(OR}_3\text{)}_2 + \text{RSH} \longrightarrow R_1R_2\text{C(SR)}_2 + R_3\text{OH}$$

常用的催化剂有氯化硅(silica chloride)[42]、ZrCl$_4$[43]、浓硫酸、对甲苯磺酸、三氟化硼/乙醚、路易斯酸、有机铝化物、LPDE 等[4,13,24,44~49]。

八、炔烃与硫醇加成

炔烃与硫醇加成可以制备缩硫醛(酮)[4]。

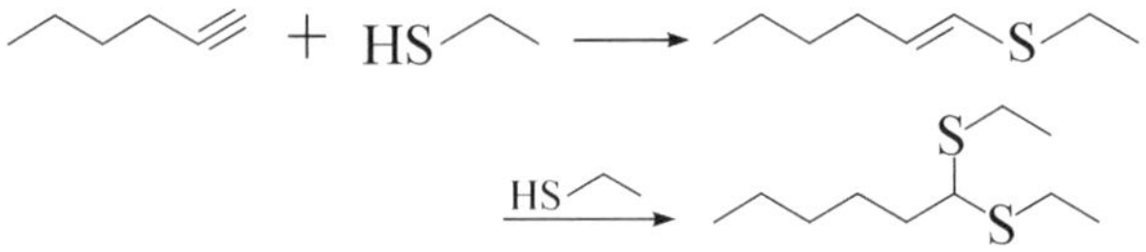

九、α,α-二卤化物与硫醇反应

α,α-二卤化物与硫醇反应可以制备缩硫醛(酮)[4,50~52],如二氯甲烷与苯硫酚反应生成二苯硫缩甲醛[53]。

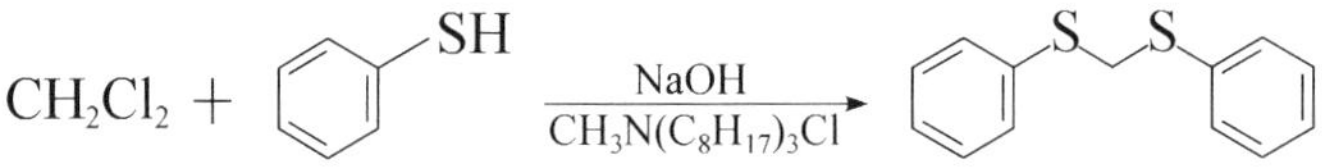

该反应可以在甲基三辛基氯化铵等相转移催化剂作用下进行[53]。

十、不对称缩硫醛(酮)的制备

醛(酮)先与六甲基二硅氮烷、硫醇反应得到中间产物,再与另一种硫醇反应,得到不对称缩硫醛(酮)。例如[16]:

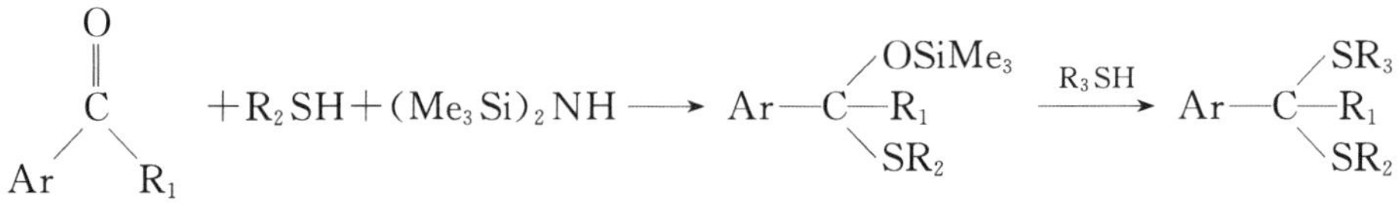

第二节 二硫代缩醛类香料

一、二甲硫基甲烷

二甲硫基甲烷[bis(methylthio)methane,FEMA 号 3878]又称二甲硫醇缩甲醛或 2,4-二硫杂戊烷,为无色透明液体,天然存在于奶酪、罐头牛肉[1]中,主要用于鸡蛋、肉味、焙烤、脂肪等食品香精。

二甲硫基甲烷可以由甲硫醇与二碘甲烷反应制备。

$$CH_3SH + CH_2I_2 \longrightarrow \text{S}\wedge\text{S} + HI$$

二、1,3-二噻烷(1,3-dithiolane)

1,3-二噻烷可以由1,3-丙二硫醇与二甲缩甲醛反应制备[54]。

$$HSCH_2CH_2CH_2SH + H_2C(OCH_3)_2 \longrightarrow \text{(1,3-dithiane)} + CH_3OH$$

1,3-二噻烷在有机合成上可用于醛酮的合成[54]。例如：

$$\text{dithiane} \xrightarrow[2) Br(CH_2)_3Cl]{1) C_4H_9Li} \text{dithiane-(CH_2)_3Cl} \xrightarrow{C_4H_9Li} \text{spiro-dithiane} \xrightarrow[HgCl_2]{H_2O} \text{cyclobutanone}$$

三、2-甲基-1,3-二噻烷

2-甲基-1,3-二噻烷(2-methyl-1,3-dithiolane，FEMA 号 3705)，又称 1,3-丙二硫醇缩乙醛，为无色至浅黄色液体，主要用于肉味、奶酪、咖啡、茶、鸡蛋、坚果等食品香精。

2-甲基-1,3-二噻烷可以由1,3-丙二硫醇与乙醛反应制备。

$$HSCH_2CH_2CH_2SH + CH_3CHO \longrightarrow \text{2-methyl-1,3-dithiane} + H_2O$$

部分二硫代缩醛类香料的化学结构和香味特征见表 8-1。

表 8-1 部分二硫代缩醛类香料的化学结构和香味特征

名称	化学结构式	香味特征
苯甲醛二甲硫醇缩醛	Ph-CH(SCH_3)_2	烤花生香气；花生香、可可粉香味[55]
苯乙醛二甲硫醇缩醛	Ph-CH_2-CH(SCH_3)_2	烤花生香气[55]
2-甲基-2-苯基乙醛二甲硫醇缩醛	Ph-C(CH_3)H-CH(SCH_3)_2	豆类、花生、炒蔬菜、水解植物蛋白香味[55]
D-葡萄糖二甲硫醇缩醛	$\begin{array}{c} H \\ CH_3S{-}{-}SCH_3 \\ H{-}{-}OH \\ HO{-}{-}H \\ H{-}{-}OH \\ H{-}{-}OH \\ CH_2OH \end{array}$	肉香[56]

续表

名称	化学结构式	香味特征
D-半乳糖二甲硫醇缩醛	$\begin{array}{c} H \\ CH_3S\!\!-\!\!\!-\!\!SCH_3 \\ H\!\!-\!\!\!-\!\!OH \\ HO\!\!-\!\!\!-\!\!H \\ HO\!\!-\!\!\!-\!\!H \\ H\!\!-\!\!\!-\!\!OH \\ CH_2OH \end{array}$	肉香、蔬菜香[56]
D-甘露糖二甲硫醇缩醛	$\begin{array}{c} H \\ CH_3S\!\!-\!\!\!-\!\!SCH_3 \\ HO\!\!-\!\!\!-\!\!H \\ HO\!\!-\!\!\!-\!\!H \\ H\!\!-\!\!\!-\!\!OH \\ H\!\!-\!\!\!-\!\!OH \\ CH_2OH \end{array}$	洋葱香、肉香[56]

第三节 1,3-二硫杂环戊烷类香料

1,3-二硫杂环戊烷类化合物是环状的硫缩醛(酮),由于其合成方法和性质有许多独特之处,因此单独列为一节进行介绍。1,3-二硫杂环戊烷类化合物在有机合成上可用作羰基的保护基[32,57,58],也常用作羰基官能团转化为双亲化合物的有机中间体[59~61]。1,3-二硫杂环戊烷类化合物作为酰基化碳负离子等价物在有机合成中起重要的作用[62~66]。由于硫对许多溶剂没有大的吸引力,所以从羰基到硫缩醛和硫缩酮的转化是许多有机合成转化的重要组成部分[67~70]。

在肉汤挥发性香成分中发现了1,3-二硫杂环戊烷、2-甲基-1,3-二硫杂环戊烷、2-丙基-1,3-二硫杂环戊烷等1,3-二硫杂环戊烷类化合物[12,71~74]。1,3-二硫杂环戊烷类化合物在咸味香精中的作用已经引起了人们的极大兴趣。

一、1,3-二硫杂环戊烷类化合物的一般合成方法

1. 醛(酮)与邻二硫醇反应

$$\begin{array}{c} R_1 \\ R_2 \end{array}\!\!\!\!\!\begin{array}{c} SH \\ SH \end{array} + O\!=\!\!\!\begin{array}{c} R_3 \\ R_4 \end{array} \longrightarrow \begin{array}{c} R_1 \\ R_2 \end{array}\!\!\!\!\!\begin{array}{c} S \\ S \end{array}\!\!\!\!\!\begin{array}{c} R_3 \\ R_4 \end{array}$$

式中:R_1,R_2,R_3,R_4 为烷基或 H。

该反应中常用的催化剂有质子酸[44,75~78],氯化钨、氯化镧、氯化钛等稀有金属卤化物[17,79~82],吸附了溴化钴等的硅酸盐[32,63,83,84],对甲苯磺酸[85],金属杂多酸[75,86~89],HSZ-360[63]等。

2. 缩醛(酮)与邻二硫醇反应

缩醛(酮)与邻二硫醇反应,生成1,3-二硫杂环戊烷类化合物[76,90~93]。

$$\text{HS-CH}_2\text{CH}_2\text{-SH} + (\text{CH}_3\text{O})_2\text{CR}_1\text{R}_2 \xrightarrow{\text{ClCH}_2\text{CH}_2\text{Cl}} \text{1,3-二硫杂环戊烷-2,2-R}_1\text{R}_2$$

3. 邻二硫醇钠和同碳二卤代烃反应

邻二硫醇钠和同碳二卤代烃缩合生成1,3-二硫杂环戊烷类化合物。例如[94]:

$$\text{NaS-CH}_2\text{CH}_2\text{-SNa} + \text{Cl}_2\text{CR}_1\text{R}_2 \longrightarrow \text{1,3-二硫杂环戊烷-2,2-R}_1\text{R}_2$$

4. 醛(酮)或缩醛(酮)与含硫有机锡化物反应

醛(酮)或缩醛(酮)与1,1-二丁基2,5-二硫杂环戊锡烷反应可以制备1,3-二硫杂环戊烷类化合物。例如[95]:

$$\text{2,2-Bu}_2\text{-1,3-二硫杂-2-锡杂环戊烷} + \text{O=CR}_1\text{R}_2 \xrightarrow{\text{Bu}_2\text{Sn(OTf)}_2} \text{1,3-二硫杂环戊烷-2,2-R}_1\text{R}_2$$

5. 醛(酮)与2-乙氧基-1,3-二硫杂环戊烷反应

醛(酮)与2-乙氧基-1,3-二硫杂环戊烷类化合物反应可以制备1,3-二硫杂环戊烷类化合物。例如[96]:

$$\text{2-乙氧基-1,3-二硫杂环戊烷} + \text{O=CR}_1\text{R}_2 \xrightarrow{\text{Bu}_2\text{Sn(OTf)}_2} \text{1,3-二硫杂环戊烷-2,2-R}_1\text{R}_2$$

6. 醛(酮)与有机硅硫化物反应

醛(酮)与二(三甲基硅)-1,2-乙二硫醚反应可以制备1,3-二硫杂环戊烷类化合物。例如[3]:

$$(\text{CH}_3)_3\text{SiSCH}_2\text{CH}_2\text{SSi}(\text{CH}_3)_3 + \text{O=CR}_1\text{R}_2 \longrightarrow \text{1,3-二硫杂环戊烷-2,2-R}_1\text{R}_2$$

7. 醛(酮)与有机硼硫化物反应

醛(酮)与1-烷基-2,5-二硫杂环戊硼烷反应可以制备1,3-二硫杂环戊烷类化合物。例如[3]:

$$\begin{matrix} S \\ | \\ S \end{matrix}B-R_3 + O=\begin{matrix}R_1\\R_2\end{matrix} \longrightarrow \begin{matrix}S\\|\\S\end{matrix}\begin{matrix}R_1\\R_2\end{matrix}$$

二、1,3-二硫杂环戊烷类香料化合物

部分1,3-二硫杂环戊烷类香料的化学结构和香味特征见表8-2。

表8-2 部分1,3-二硫杂环戊烷香料的化学结构和香味特征

名称	化学结构式	香味特征
1,3-二硫杂环戊烷		洋葱香、烤大蒜牛肉香气
2-甲基-1,3-二硫杂环戊烷		烤洋葱香、蔬菜香气、烤肉香[97]
4-甲基-1,3-二硫杂环戊烷		洋葱香、类菜根的味道、牛肉香
2,2-二甲基-1,3-二硫杂环戊烷		肉香
2,4-二甲基-1,3-二硫杂环戊烷		类似洋葱、轻金属特征的菜香气
4-甲基-2-甲硫基甲基-1,3-二硫杂环戊烷		肉香、HVP样香气;洋葱香、烤洋葱香、HVP样味道[98]
4,5-二甲基-2-甲硫基甲基-1,3-二硫杂环戊烷		甜的、肉香、HVP样香气;甜的、肉香、HVP样、面包味道[98]
4-甲基-2-甲硫基乙基-1,3-二硫杂环戊烷		蘑菇香、HVP香、烤洋葱香、肉香[98]
4,5-二甲基-2-甲硫基乙基-1,3-二硫杂环戊烷		肉香,HVP香、火鸡肉香、鸡肉香、猪肉香味[98]
4-甲基-2-甲硫基丙基-1,3-二硫杂环戊烷		蘑菇香气,肉香味
4,5-二甲基-2-甲硫基丙基-1,3-二硫杂环戊烷		类蔬菜香气,肉香味
2-乙基-1,3-二硫杂环戊烷		洋葱香气,肉香

续表

名称	化学结构式	香味特征
2-正丙基-1,3-二硫杂环戊烷		花香,新鲜的肉汤香气
2-异丙基-1,3-二硫杂环戊烷		葱蒜香气,辛香
2-丙烯基-1,3-二硫杂环戊烷		烤肉香气,稍甜
2-正丁基-1,3-二硫杂环戊烷		大蒜香气,稍带有刺激味
2-异丁基-1,3-二硫杂环戊烷		葱香、蒜香,烤肉香,金属样香气[99]
2-正己基-1,3-二硫杂环戊烷		辛香,类似洋葱香气,肉香味
2-正庚基-1,3-二硫杂环戊烷		类似葱蒜香气,稍带刺激性气味
2-正辛基-1,3-二硫杂环戊烷		辛香,烤肉香,蒜香
2-正壬基-1,3-二硫杂环戊烷		浓郁的葱蒜香气,肉腥味
2-(2-呋喃基)-1,3-二硫杂环戊烷		炖肉香气,稍甜
2-苯基-1,3-二硫杂环戊烷		烤肉香,稍苦
2,6-二甲基-1-(1,3-二硫杂-2-环戊烷基)-1,5-庚二烯		淡淡的清香,稍带甜味
2-甲基-2-乙基-1,3-二硫杂环戊烷		微弱的肉香,带臭皮蛋气息[100]

续表

名称	化学结构式	香味特征
2-甲基-2-丙基-1,3-二硫杂环戊烷		微弱的热带水果香气,似菠萝、西番莲香气[100]
2-甲基-2-丁基-1,3-二硫杂环戊烷		微弱的热带水果香气,似菠萝、西番莲香气[99]
2-甲基-2-异丁基-1,3-二硫杂环戊烷		葱蒜香气[101]
2-甲基-2-异戊基-1,3-二硫杂环戊烷		葱香香气,稍带甜味[101]
1,4-二硫杂螺[4,4]壬烷		微弱的葱蒜香气[99]
1,4-二硫杂螺[4,5]癸烷		微弱的油脂香气,肉香味[100]
6-甲基-1,4-二硫螺[4,5]癸烷		大蒜香气,稍苦,肉味
2-甲基-2-异丁烯基-1,3-二硫杂环戊烷		微弱的热带水果香气[100]
1-(2-甲基-1,3-二硫杂-2-环戊烷基)丙酮		肉香,萝卜干样香气[100]
1-(2-甲基-1,3-二硫杂-2-环戊烷基)乙酸乙酯		优美的果香,似热带水果香气[99]
1,3-二(1,3-二硫杂-2-环戊烷基)丙烷		微弱的果香,肉香味[99]
2,4-二甲基-2-正丁基-1,3-二硫杂环戊烷		葱蒜香气,并有些刺激性气味,焦味

第四节　其他二硫代缩酮类香料

二硫代缩酮类香料化合物目前批准允许作为食品香料使用的品种很少,但这类化合物是一类潜在的很有发展前途的香料。其他一些二硫代缩酮类香料的化学结构和香味特征见表8-3。

表8-3 其他一些二硫代缩酮类香料的化学结构和香味特征

名称	化学结构式	香味特征
苯乙酮二甲硫醇缩酮		洋葱香、榴莲香、热带水果香、乳酪香、硫化物香气[55]
丁酮二正丁硫醇缩酮		带油香的葱蒜香[34]
环己酮二正丁硫醇缩酮		强烈葱蒜香[35]
螺[2,4-二硫杂-1-甲基-8-氧杂二环[3.3.0]辛烷-3,3'-(1'-氧杂-2'-甲基)环戊烷]		肉香味
螺[2,4-二硫杂-6-甲基-7-氧杂二环[3.3.0]辛烷-3,3'-(1'-氧杂-2'-甲基)环戊烷]		肉香味

参 考 文 献

[1] Morton I D, Macleod A J. Food Flavours. Part A. Introduction. New York: Elsevier Scientific Publishing Company, 1982. 241~242

[2] Graham A E. A simple and convenient catalytic procedure for the preparation of dithioacetals. Synth. Commun., 1999, 29(4): 697~703

[3] 西奥多拉·W. 格林. 有机合成中的保护基. 1981. 范如霖译. 上海:上海科技文献出版社, 1985. 152~166

[4] Reid E E. Organic Chemistry of Bivalent Sulfur. New York: Chemical Publishing Co. Inc., 1960. 320~329

[5] Kuliev A M. Synthesis and study of the effect of some mercaptals on the properties of motor oils. Prissdki Smaz. Maslam., 1967, 106~109

[6] Pittet A. Flavoring with gem dithioethers of phenylalkanes: US, 4585663. 1986.4.29

[7] Pittet A O, Courtney T F, Muralidhara R. Flavoring with phenylalkyl mercaptals: US, 4504508. 1985.3.12

[8] Pittet A. Flavoring with alkylthioalkanal dialkyl mercaptals: US. 4481224. 1984.11.6

[9] Pittet A. Flavoring with dialkylthioalkanes, dialkyl thioalkylcycloalkenes and monoalkyl

thioalkenylcycloalkenes: US, 4565707. 1986.1.21

[10] Pittet A. Flavoring with furfuryl mercaptals: US, 4515429. 1985.4.30

[11] Pittet A O, Vock M H, Courtney T F et al. Uses of methyl substituted 2-(2,6-dimethyl-1,5-heptadienyl)-1,3-dithiolanes in augmenting or enhancing the aroma or taste of foodstuff: US, 4464408. 1983.8.7

[12] Pittet A. Uses of methyl(methylthioalkyl)-1,3-dithiolanes augmenting or enhancing the aroma or taste of foodstuffs: US, 4472446. 1984.9.18

[13] Lepez Aparricio F J, Zorrilia Benitez F, Santoyo Gonzalez F. Synthesis and structures of some diglycoladehyde thioacetals. Carbohydr. Res., 1982, 102(1): 69~81

[14] Мусавиров Р С, Газизова Л Б, Недогрей Е П. Получениеиреакции 1,1-ди(этилтио)алканов. Журнал Общей Химии, 1981, 51(5): 1066~1070

[15] Lopez Aparico F J, Zorrilia Benitez F, Santoyo Gonzalez F. Acetals and thioacetals from thiodiglycolaldehyde: some oxidation froducts. Carbohydr. Res., 1982, 110(2): 195~205

[16] Amato J S. Process for unsymmetrical dithioacetals and dithioketals: EP, 328188. 1989.8.16

[17] Kumar V, Sukh D. Titanium tetrachloride, an efficient and convenient reagent for thioacetalization. Tetrahedron Letter, 1983, 24(12): 1289~1292

[18] Cohn T, Kosarych Z. Preparative methods for (Z)-2-methoyy-1-phenylthio-1,3-butadienes. Tetrahedron, 1980, 12(41): 3955~3958

[19] Ong B S. Synthesis, intramolecular charge-tranfer interaction, and electron transport properties of 9,9-bis(alkylthio)nitrofluorenes. J. Chem. Soc. Chem. Commun., 1984, (5): 266~268

[20] 张克强, 孙宝国. 苯甲醛正丁硫醇缩醛的合成研究. 精细化工, 1998, 15(3): 23~25

[21] 张克强, 孙宝国. 柠檬醛甲硫醇缩醛的合成研究. 精细化工, 1998, 15(2): 25~26

[22] 付翔, 孙宝国. 乙醛正丁硫醇缩硫醛的合成. 精细化工, 2000, 17(7): 384~385

[23] Firouzabadi H, Iranpoor N, Karimi B. Lithium bromide-catalyzed highly chemoselective and efficient dithioacetalization of α,β-unsaturated and aromatic aldehydes under solvent-free conditions. Synthesis, 1999, (1): 58~60

[24] Park J H, Kim S. Magnesium bromide mediated selective conversion of acetals into thioacetals. Chem. Lett., 1989, (4): 629~632

[25] Labiad B, Villemin D. Clay catalysis: Synthesis of organosulphur synthons. Synth. Comun., 1989, 19(1~2): 31~38

[26] 张克强, 孙宝国. 糠醛异丙硫醇缩醛的合成研究. 精细化工, 1998, 15(1): 30~31

[27] 张克强, 孙宝国. 1-(并二甲硫基)甲基-4-甲氧基苯的合成. 精细化工, 1997, 14(4): 25~27

[28] 张克强, 孙宝国. 茴香醛二烷基缩硫醛的合成研究. 精细化工, 1999, 16(增): 277~279

[29] Firouzabadi H, Karimi B, Eslami S. Lithium trifluoromethanesulfonate (LiOTf) as a highly efficient catalyst for chemoselective dithioacetalization of carbonyl compounds under

neutral and solvent-free conditions. Tetrahedron Lett., 1999, 40(21): 4055~4058

[30] Pemi R B. Amberlyst-15 as a convenient catalyst for chemoselective thioacetalization. Synth. Comun., 1989, 19(13~14): 2383~2387

[31] Kakimoto M A, Seri T, Imai Y. Synthesis of dithioacetals from carbonyl compounds and thiols in the presence of polyphosphoric acid trimethylsiyl ester. Synth., 1987, (2): 164~165

[32] Kamitori Y, Hojo M. Selective protection of carbonyl compounds. Silica gel treated with thionyl chloride as an effective catalyst for thioacetalization. J. Org. Chem., 1986, 51: 1427~1431

[33] Tazaki M, Takagi M. Deoxygenative thioacetalization of carbonyl compounds with organic disulfide and tributylphosphine. Chem. Lett., 1979, (7): 767~770

[34] 余爱农,蔡传斌.丁酮二正丁硫醇缩酮的合成研究.精细化工,1999,16(增):259~260

[35] 吕银华,余爱农,刘江燕.环己酮二正丁硫醇缩酮合成的研究.精细化工,1999,16(增):301~302

[36] Cama L D. Total synthesis of thienamycin analogs-Ⅲ. Syhthesis of 2-aryl and 2-heteroaryl analogs of thienamycin. Tetra., 1983, 39(15): 2531~2549

[37] Kim T H, Oh D Y. Synthesis of S,S-thioacetals of formyphosphonate from chloro(arythio)methanephosphonate. Synth. Comun., 1988, 18(14): 1611~1614

[38] Mikolajczwk M et al. Tetrahedron Letter, 1998, 30(20): 3081~3088

[39] 张克强.缩硫醛类化合物的合成与研究[硕士学位论文].北京:北京轻工业学院,1997. 11~12

[40] Bestmann H J, Engler R, Hartung H et al. Reaktionen von phosphinalkylenen mit schwefelkohlenstoff — Synthese von ketenmercaptalen und dithiocarbonsäureestern. Chem. Ber., 1979, 112(11): 28~41

[41] Denis J N, Desauvage S, Hevesi L et al. New synthetic routs to vinyl sulfides, ketene thioacetals and their seleno analogues from carbonyl compounds. Tetrahedron Lett., 1981, 22(40): 4009~4012

[42] Firouzabadi H, Iranpoor N, Karimi B et al. Highly efficient transdithioacetalization of acetals catalyzed by silica chloride. SYNLETT, 2000, (2): 263~265

[43] Firouzabadi H, Iranpoor N, Karimi B. Zirconium tetrachloride ($ZrCl_4$) catalyzed highly chemoselective and efficient transthioacetalization of acetals and acetalization of carbonyl compounds. SYNLETT, 1999, (3): 319~323

[44] Saraswathy V G, Sankaraman S. Chemoselective protection of aldehydes as dithioacetals in lithium perchlorate-diethyl ether medium. Evidence for the formation of oxocarbenium ion intermediate from acetals. J. Org. Chem., 1994, 59(16): 4665~4670

[45] Cohen T. Base-induced ring closure of bis-dithioacetals via an apparent carbenoid. A Nobel route to functionalized cyclobutanes and cyclopentanes. Tetrahedron Lett., 1978, (51): 5063~5066

[46] Satoh T, Uwaya S, Yamakawa K. Bis (diisobutylaluminum) 1,2-ethanedithiolate: a reagent in the conversion of oxyacetals to thioacetals. Chem. Lett., 1983, (5): 667~670

[47] Mssaki Y, Serizawa Y, Kaji K. Monothioacealization of acetals using diethylaluminium thiophenoxide. Chem. Lett., 1985, (12): 1933~1936

[48] Kim S, Park J H, Lee S. Conversion of actals into monothioacetals, α-alkoxyazides and α-alkoxyalkyl thioacetates with magnesium bromide. Tetrahedron Lett., 1989, 30(48): 6697~6700

[49] Tanimoto S, Iwata S, Imanishi T et al. Interchange reactions of diphenyl acetals and alkyl dialkylaminomethyl ethers with thiols. Bull. Inst. Chem. Res., Kyoto Univ. 1979, 57 (3): 235~239

[50] Kumer S J. Timber Dev Assoc India, 1985, 31(2): 18~26

[51] Crestan H J, Chabaud B, Labaudiniere R et al. Synthesis of vinyl selenides or sulfides and ketene selenoacetals or thioacetals by nickel(Ⅱ) vinylation of sodium benzeneselenolate or benzenethiolate. J. Org. Chem., 1986, 51: 875~878

[52] 孙宝国，何坚. 香料化学与工艺学. 第二版. 北京：化学工业出版社，2004. 521~522

[53] Herriott W. The Phase-transfer synthesis of sulfides and dithioacetals. Synthesis, 1975, (7): 447~448

[54] 宗汝实. 极性转换在有机合成中的应用. 有机化学, 1981, (5): 323~335

[55] 张克强, 孙宝国. 缩硫醛(酮)类化合物的合成及应用. 精细化工, 1999, 16(增): 283~290

[56] Oftedahl M L. Flavoring food with a sugar dialkyl dithioacetals: US, 3635736. 1972. 1.18

[57] Kumar P, Reddy R S, Zeolite H Y. An efficient catalyst for thioacetalizayion. Tetrahedron Letter., 1992, 33 (6):825~826

[58] Wilson G E, Huang M G. Facile synthesis of 1,3-dithiolanes from ketone and 2-mercaptoethanol. J. Org. Chem., 1968, 33(5): 2133~2144

[59] Larson G L, Hernandez A. Reactions of trimethylsilyl enol ethers with 2-mercaptoethanol. Synthetic Communication, 1974, 4(1): 61~62

[60] Tanimoto S, Miyakes T, Okano M. Synthesis of 2-ethoxy-1,3-oxathiolane and 2-ethoxy-1,3-dithiolane and their some reactions. Bull. Inst. Chem. Res., Kyoto Univ. 1977, 55 (3): 276~281

[61] Sabde D P, Naik B G, Hegde V R et al. Dithioacetalization of carbonyl compounds and tetrahydropyranylation of alcohols over H-Rho zeolite. J. Chem. Research, 1996, (11): 494~495

[62] Keskinen R, Nikkila A, Pihlaja K. Conformational analysis, Part VII: 2-alkyl-4-methyl- and 2-alkyl-2, 4-dimethyl-1, 3-dithiolanes. J. Chem. Soc., Perkin II. 1973, 12 (1): 1376~1379

[63] Ballini R, Barboni L. Thioacetalization of carbonyl compounds by zeolite HSE-360 as a new, effective heterogeneous catalyst. Synthetic Communications., 1999, 29 (5):

767~772

[64] Reid E E, Jelinek A. Some mercaptoles of 1,2-ethanedithiol. SYNLETT, 1967, (2): 448~449

[65] Viski P, Waller D P. Strutures of the intermediates of metal-ion-promoted ring opening reactions of 1,3-dithiolanes and 1,3-dithianes in nonaqueous media. J. Org. Chem., 1996, 61: 7631~7632

[66] Chou W C, Yang S A, Fang J M. Reactions of 1,3-dithiolaneand 1,3-dioxides with nucleophiles. J. Chem. Soc., Perkin Trans I. 1994, (5): 603~609

[67] Bellesia F, Boni M, Ghelfi F. Ketene Thioacetals from chloromercaptals. Synthetic Communications, 1993, 23(22): 3179~3184

[68] Colonna S, Gaggero N. Enantio and diastereoselectivity of cyclohexanone monooxygenase catalyzed oxidation of 1,3-dithioacetals. Tetrahedron, 1996, 7(2): 565~570

[69] Ku B, Oh D Y. Tetrachlorosilane catalyzed dithioacetalization. Synthetic Communication, 1989, 19(3,4): 433~438

[70] Choudary B M, Sudha Y. Chemoselective thioacetalisation of aldehydes catalysed by Fe^{3+}-montmorillonite^{+}. Synthetic communication, 1996, 26(16): 2993~2997

[71] Wilson R A, Pascale J V. Six membered heterocyclic oxathio compounds: US, 4042601. 1977.8.16

[72] Richard A, Edison W, Mussinan C. Flavoring and aromatising with a five or six mumbered heterocyclic dithio compounds: US, 3863 013. 1972.1.28

[73] Mikhailova T V, Gren A I. Identification of sulphur-organic compounds obtained by thermal treatment of the meat broths in the prence of alkyl-mercaptopropanol. Die Nahrung, 1985, 29(7): 671~680

[74] Garbusov V, Rehfeld G. Volatile sulfur compounds contributing to meat flavor. Part I. Components identified in boiled meat. Die Nahrung, 1976, 20(3): 235~241

[75] Curini M, Epifano F. An efficient procedure for the preparation of cyclic ketals and thioketals catalyzed by zirconium sulfophenyl phosphonate. SYNLETT, 2001, (7): 1182~1184

[76] Firouzabadi H, Iranpoor N. Trichloroisocyanuric acid as a mild and efficient catalyst for thioacetalization and transthio-acetalization reactions. SYNLETT, 2001, (10): 1641~1643

[77] Ballini R, Bosica G. Amberlyst 15 as a mild, chemoselective and reusable heterogeneous catalyst for the conversion of carbonyl compounds to 1,3-oxathiolanes. Synthesis, 2001, (12): 1826~1829

[78] Kobayashi S, Iimura S. Dehydration reactions in water. Surfactant-type bronsted acid-catalyzed dehydrative etherification, thioetherification, and dithioacetalization inwater. Chemistry Lett., 2002, 8(1): 10~11

[79] Garlaschelli L, Vidari G. Anhydrous lanthanum trichloride, a mild and convenient reagent for thioacetalization. Tetrahedron Lett., 1998, 30(40): 5815~5816

[80] Firouzabadi H. Tungsten hexachloride (WCl_6) as an efficient catalyst for chemoselective

dithioacetalization of carbonyl compounds and transthioacetalization of acetals. Nasser Iranpoor. SYNLETT. 1998, 739~740

[81] Zhang Y M, Yu Y P. Samarium triiodide catalyzed dithioacetal and dithioketal formation. Oppi Briefs., 1993, 25(3): 365~367

[82] Patney H K. Anhydrous Cobalt bromide dispersed on silica gel: a mild and efficient reagent for thioacetalisation of carbonyl compounds. Tetrahedron Lett., 1994, 35(31): 5717~5718

[83] Patney H K. Bis(trimethylsilyl) sulfate-silica catalysed thioacetalisation of carbonyl compounds. Tetrahedron Lett., 1993, 34(44): 7127~7128

[84] Miranda R, Cervantes H. Preparation of dithiolanes by bentonitic earth catalysis. Synthetic Communication, 1990, 20(1): 153~157

[85] Streinz L, Koutek B, Saman D. An effective synthesis of 1,3-dithiolane. Collect Czech Chem., 1997, (6): 665~671

[86] Kumar P, Subhas M, Roy C. Ceric ammonium nitrate as a convenient catalyst for chemoselective thioacetalisation. Tetrahedron, 1997, 51(28): 7823~7828

[87] Anand R V, Saravanan P. Solvent free thioacetalization of carbonyl compounds catalyzed by $Cu(OTf)_2 SiO_2$. SYNLETT, 1999(4): 415~416

[88] Deka N, Sarma J C. Highy efficient dithioacetalization of carbonyl compounds catalyzed with Iodine supported on neutral alumina. Chemistry Lett., 2001, (5): 794~795

[89] Yadav J S, Reddy B V, Pandey S K. $LiBF_4$ catalyzed chemoselective conversion of aldehydes to 1,3-oxathiolanes and 1,3-dithianes. SYNLETT, 2001, (2):238~239

[90] Firouzabadi H, Iranpoor N. Iodine catalyzes efficient and chemoselective thioacetalization of carbonyl functions, transthioacetalization of O,O-and S,O-acetals and acylals. J. Org. Chem., 2001, 66: 7527~7529

[91] Ranu B C, Das A. Highly efficient transthioacetalization of O,O-acetals catalyzed by indium chloride. SYNLETT, 2002, 5(03): 727~730

[92] Jnaneshwara G K, Wakharkar R D. Transdithioacetalization of acetals, ketals, oxims, enamines and tosylhydrazones catalysed by natural kaolinitic clay. J. Chem. Soc., Perkin Trans I, 1998, 45: 965~968

[93] Gajare A S, Shingare M S, Bandgar B P. Selective transdithioacetalization of acetals, ketals, oxathioacetals and oxathioketals catalyzed by envirocat $EPZ10^R$. J. Chem. Research, 1998, (8): 452~453

[94] Jones R H, Lukes G E et al. Insecticidal substituted 1,3-dithiolanes and method of application: US,2690988. 1954.10.5

[95] Sato T, Otera J, Histosi Nozaki. Differentiation between carbonyls and acetals in 1,3-dithiane and 1,3-dithiolane synthesis catalyzed by organotin triflates. J. Org. Chemistry, 1993, 58:4971~4978

[96] Jo S, Oida T. The reaction of 2-ethoxy-1,3-dithiolane with carbonyl compounds. Bull.

Chem. Soc. Japan, 1981, 54: 1434~1436

[97] Macleod G. The flavor of beef//Shahidi F. Flavor of Meat, Meat Products and Seafoods. London: Blackie Academic & Professional, 1998. 28~60

[98] Pittet A O, Courtney T F, Muralidhara R. Methyl(methylthioalkyl)-1,3-dithiolanes: US, 4515967. 1985.5.7

[99] 郑福平,李海军,孙宝国等. 磷钨酸催化合成1,3-二硫杂环戊烷类香料化合物. 中国食品学报, 2006, 6(1): 63~66

[100] 郑福平,李海军,孙宝国等. 对甲苯磺酸催化合成1,3-二硫杂环戊烷类香料化合物. 食品科技, 2006,(1):79~91

[101] 李海军,郑福平,孙宝国等. 3种2-甲基-2-烷基-1,3-二硫杂环戊烷类化合物的合成. 精细化工, 2005,22(3):195~197

第九章 硫代羧酸酯类香料

硫代羧酸酯类化合物是咖啡[1]、芝麻油[2,3]等食物的挥发性香成分。在大部分热反应模型体系的香成分中都发现了硫代羧酸酯类化合物,如 D-葡萄糖、氨、硫化氢模型体系[4,5]和 D-葡萄糖、氨、硫化氢、半胱氨酸模型体系[6]中都发现了硫代糠酸甲酯。硫代羧酸酯类化合物在肉味、海鲜、咖啡、芝麻油、奶制品等香精中具有重要应用。

硫代糠酸酯类化合物具有抗菌作用,其中硫代糠酸酚酯类化合物的抗菌效果更好[7],如硫代糠酸 3,4-二氯苄酯可用作土壤杀菌剂[8],硫代糠酸酯类化合物也可用作杀螨剂[9]。

硫代羧酸酯也是有机合成的重要中间体,如硫代羧酸酯与格氏试剂反应可以制备不对称酮。

$$R_1C(O)S\text{-}Py + R_2MgBr \longrightarrow R_1C(O)R_2 + Py\text{-}SMgBr$$

此方法已经成功地用于重要香料顺-茉莉酮的合成[10]。

第一节 硫代羧酸酯类香料的一般合成方法

一、酰卤与硫醇或其钠盐反应

酰卤与硫醇或其钠盐反应可以制备相应的硫代羧酸酯[11~14]。例如,乙酰卤与乙硫醇反应可以制备硫代乙酸乙酯[15]。

$$CH_3C(O)X + HSC_2H_5 \longrightarrow CH_3C(O)SC_2H_5 + HX$$

式中:X 为 Cl 或 Br。

二、酸酐与硫醇反应

酸酐与硫醇反应很容易生成相应的硫代羧酸酯[11,13]。例如,糠硫醇与乙酸酐反应可以制备硫代乙酸糠酯[16]。

$$\text{furyl-CH}_2\text{SH} + (CH_3CO)_2O \xrightarrow{CH_3COONa} \text{furyl-CH}_2\text{-S-CO-CH}_3 + CH_3COOH$$

三、酸酐与三硫烷基硅烷反应

酸酐与三硫烷基硅烷混合后在催化剂作用下反应生成相应的硫代羧酸酯。例如：

$$(R_1CO)_2O + HSi(SR)_3 \longrightarrow R_1\text{-CO-SR}$$

四、酸酐或酰氯与 XMgSR 反应

硫代羧酸酯可以很容易地由 XMgSR 与酰氯或酸酐反应而制取。反应分两步进行，第一步为硫醇与格氏试剂反应，所得的中间产物再与相应的酰氯或酸酐反应得到最终产物硫代羧酸酯。例如：

$$R_1SH + R_2MgX \longrightarrow R_1SMgX$$

$$R_1SMgX + R_3\text{-CO-Cl} \longrightarrow R_3\text{-CO-SR}_1$$

五、羧酸与硫氰酸酯反应

羧酸或酰氯与硫氰酸酯在催化剂作用下反应生成相应的硫代羧酸酯。例如：

$$R_1\text{-CO-X} + R_2SCN \longrightarrow R_1\text{-CO-SR}_2$$

式中：X 为 OH 或 Cl。

六、硫代羧酸(盐)与磺酸酯反应

硫代羧酸(盐)与磺酸酯反应生成硫代羧酸酯。例如[17]：

$$\text{4-CH}_3\text{-C}_6\text{H}_4\text{-SO}_3\text{C}_3\text{H}_7 + CH_3\text{-CO-SK} \longrightarrow CH_3\text{-CO-S-C}_3\text{H}_7$$

七、硫代羧酸(盐)与卤代烃反应

硫代羧酸(盐)与卤代烃反应生成相应的硫代羧酸酯。例如[18]：

$$\text{CH}_3\text{COSK} + \text{（3-氯-2-乙基呋喃）} \xrightarrow{N,N\text{-二甲基甲酰胺}} \text{产物}$$

八、硫代羧酸与烯烃加成

硫代羧酸与不饱和烯烃化合物反应生成硫代羧酸酯类化合物。例如[19]：

$$\text{环己烯} + \text{CH}_3\text{COSH} \longrightarrow \text{环己基硫代乙酸酯}$$

第二节 硫代羧酸酯类香料

一、硫代甲酸糠酯

硫代甲酸糠酯(furfuryl thiolformate, FEMA 号 3158)为无色透明液体，具有烤肉香、坚果香、咖啡香，可用于肉味、咖啡、坚果、奶制品等食品香精，在最终加香食品中浓度约为 1mg/kg。

硫代甲酸糠酯可以由甲乙酸酐与糠硫醇反应制备[11,20]。

$$\text{CH}_3\text{COOCOH} + \text{糠基-CH}_2\text{SH} \longrightarrow \text{糠基-CH}_2\text{SCHO} + \text{CH}_3\text{COOH}$$

二、硫代糠酸甲酯

硫代糠酸甲酯(methyl-2-thiofuroate, FEMA 号 3311)为无色至浅黄色液体，天然存在于咖啡、香油等香成分中，具有肉香、海鲜香、甘蓝香味，可用于肉类、咖啡、奶制品等食品香精。在最终加香食品中浓度为 3～5mg/kg。

硫代糠酸甲酯可以由糠酰氯与甲硫醇反应制取[21,22]。

$$\text{糠酰氯} + \text{CH}_3\text{SH} \longrightarrow \text{硫代糠酸甲酯} + \text{HCl}$$

三、2,5-二甲基-3-乙酰硫基呋喃

2,5-二甲基-3-乙酰硫基呋喃（2,5-dimethyl-3-acetylthiofuran, FEMA 号 4034)具有烤肉香[23]，是肉味香精的关键香料之一，可以通过下面的路线合成[23]：

四、其他硫代羧酸酯香料

其他硫代羧酸酯类香料的化学结构和香味特征见表 9-1。

表 9-1 其他硫代羧酸酯类香料的化学结构和香味特征

名称	化学结构式	香味特征或应用建议
硫代乙酸甲酯 (FEMA 号 3876)		强烈的洋葱、大蒜、萝卜样香味[24]
硫代乙酸乙酯 (FEMA 号 3282)		洋葱香、大蒜香、果香、肉香香气;洋葱、大蒜味道[25]
硫代乙酸丙酯 (FEMA 号 3385)		青香、洋葱香、大蒜香、蔬菜香气和味道[25]
硫代乙酸糠酯 (FEMA 号 3162)		奶酪香、烤香、葱蒜香、蔬菜香、咖啡香气;奶酪香、葱蒜香、咖啡香、肉香味道[11]
硫代丙酸甲酯 (FEMA 号 4172)		可用于焙烤食品、饮料、口香糖、奶制品、调味品等香精[26]
硫代丙酸糠酯 (FEMA 号 3347)		咖啡、洋葱、大蒜香气;咖啡、洋葱、大蒜、肉香味道[11]
硫代丁酸糠酯		咖啡香[11]
硫代 3-甲基-2-丁烯酸糠酯		咖啡香、焦糖香、坚果香[11]
硫代惕各酸糠酯		咖啡香、蘑菇香味[11]
2-甲基-3-乙酰硫基呋喃 (FEMA 号 3973)		肉香味[27, 28]
硫代丙酸烯丙酯 (FEMA 号 3329)		洋葱、大蒜、青香香气;洋葱、青香、甜的、蔬菜味道[25]

续表

名称	化学结构式	香味特征或应用建议
硫代丁酸甲酯 (FEMA 号 3310)		奶酪香、西红柿香、霉香、葱蒜香气；奶酪香、西红柿香、青香、热带水果味道[29]
硫代 3-甲基丁酸甲酯 (FEMA 号 3864)		热带水果香韵[30]
硫代丁酸 3-己酯 (FEMA 号 4136)		可用于饮料、口香糖、糖果、奶制品等香精[26]
硫代 3-甲基-2-丁烯酸 2-丁酯		辛香香气[30]
硫代己酸甲酯 (FEMA 号 3862)		热带水果香韵[30]
硫代己酸烯丙酯 (FEMA 号 4076)		可用于焙烤食品、饮料、口香糖、肉制品、奶制品、调味品等香精[26]
硫代苹果酸二异戊酯 (FEMA 号 4096)		可用于焙烤食品、饮料、肉制品、奶制品、调味品等香精[26]
硫代苯甲酸甲酯 (FEMA 号 3857)		霉臭味[25]
硫代糠酸乙酯		清新海鲜味；煎虾香气
硫代糠酸丙酯		烤肉香气[31]
硫代糠酸异丙酯		鲜淡海鲜香气、坚果香气[31]
硫代糠酸丁酯		烤肉香气；鸡肉香味[31]
硫代糠酸糠酯		大蒜香[11]
硫代糠酸 2,5-二甲基-3-呋喃酯 (FEMA 号 3481)		HVP 香、肉香[25]

参 考 文 献

[1] Stoll M, Winter M, Gautschi et al. 68. Recherches sur les aromes. Helv. Chim. Acta, 1967, 50(2): 628~694

[2] Nakamura S, Nishimura O, Masuda H et al. Identification of volatile flavor components of roasted sesame oil//Bhattacharyya S C, Sen N, Sethi K L. 11th International Congress of Essential Oils, Fragrances and Flavours. Proceedings: Volume5, Chemistry—Analysis, Structure and Synthesis. New Delhi: Oxford & Ibh Publishing Co. Pvt. Ltd., 1989. 73~87

[3] Nakamura S, Nishimura O, Masuda H. Identififcation of volatiole flavor components of the oil from roasted sesame seeds. Agric. Biol. Chem., 1989, 53(7): 1891~1899

[4] Takayuki S, Russell G F. Study of meat volatiles associated with aroma generated in a D-glucose-hydrogen sulfide-ammonia model system. J. Agri. Food Chem., 1976, 24(4): 843~846

[5] Takayuki S, Russell G F. A study of the volatiles isolated from a D-glucose-hydrogen sulfide-ammonia model system. J. Agri. Food Chem., 1977, 25(1): 109~112

[6] 坂口稔. Formation of heterocyclic compounds from the reaction of D-glucose and hydrogen sulfide, ammonia, cysteine or cysteamine. 香料(日), 1984, 144: 85~87

[7] 丹羽荣二, 青木博夫, 田中博等. フランカルボン酸チオエステル类の合成. 农芸化学会誌, 1967, 41(4): 135~138

[8] Oyama Hiroshi, Umeda Ten. 3,4-Dichlorobenzylthiol ester derivative and soil germicide: 日本, 公开特许公报 昭 60-149555. 1985.8.7

[9] Marie M J. Acaricidal cmpositons: EP, 0078562. 1983.05.11

[10] 沈延昌. 元素有机化合物在有机合成中的应用. 有机化学, 1985, (1): 35~43

[11] Max W, Goldman I M. flavoring agent: US, 3943260. 1976.3.9

[12] 郑福平, 孙宝国, 何坚. 硫代糠酸乙酯和硫代糠酸糠酯的合成. 精细化工, 1997, 14(5): 17~19

[13] Wenzel F W, Reid E E. The preparation and properties of a series of alkyl thiolacetates. J. Am. Chem. Soc., 1937, 59(1): 1089~1091

[14] 孙宝国, 郑福平, 刘玉平. 香料与香精. 北京:中国石化出版社, 2000: 149~150

[15] Schaefgen J R. Kinetics of the hydrolysis of ethyl thiolacetate in aqueous acetone. J. Am. Chem. Soc., 1948, 70(4): 1308~1311

[16] Adams R, Gaufbke C G. 2-Furylmethyl acetate. Org. Syn., 1956, Coll. 1: 285~286

[17] Chapman J H, Owen L N. Dithiols. Part Ⅳ. The reaction of toluene-p-sulphonates and methanesulphonates with potassium thilacetate: a new method for the preparation of thiols. J. Chem. Soc., 1950, 579~584

[18] 徐晶, 孙宝国. 2-乙基-3-乙酰硫基四氢呋喃的合成研究. 精细化工, 1999, 16(增): 368~370

[19] Cunneen J I. The addition of thio-compounds to olefins. Part Ⅱ. Reactions of thioacetic and mono-, di-, and tri-chlorothiolacetic acids. J. Chem. Soc., 1947, (1): 134~147

[20] 郑福平,孙宝国,何坚. 硫代甲酸糠酯的合成研究. 化学世界,1997, 38(6):312~313

[21] Winter M, Flament I, Stoll M et al. Flavor modified soluble coffee: US, 3702253. 1972.11.7

[22] 郑福平,孙宝国,何坚. 硫代糠酸甲酯合成方法的改进. 化学试剂,1998,20(4):248~249

[23] Evers W J, Heinsohn H R, Mayers B J. Processes for producing 3-thia furans and 3-furan thiols: US, 3922288. 1975.11.25

[24] 孙宝国,徐理阮. 硫代乙酸甲酯的合成研究. 北京轻工业学院学报,1995, 13(1): 42~45

[25] 孙宝国,何坚. 香精概论. 第二版. 北京:化学工业出版社, 2006. 126~151

[26] Smith R L, Cohen J, Doull J et al. GRAS flavoring substances 22. Food Technology, 2005, 59(8):24~62

[27] MacLeod G, Seyyedain-Ardebili M. Natural and simulated meat flavors (with particular reference to beef). CRC Critical Reviews in Food Science and Nutrition, 1981, 14(4): 308~437

[28] Rowe D J. Aroma chemicals for savory flavors. Perfumer & Flavorist, 1998, 23(4): 9~16

[29] Mosciano G et al. Organoleptic characteristics of flavor materials. Perfumer & Flavorist, 1997, 22(4): 57~59

[30] Rowe D. More fizz for your buck: High-impact aroma chemicals. Perfumer & Flavorist, 2000, 25(5): 1~19

[31] 郑福平,孙宝国,何坚. 硫代糠酸酯类香料化合物的合成. 精细化工,1995, 12(5):26~28

第十章　噻吩类香料

噻吩环上的碳原子与硫原子以 sp² 杂化轨道互相连成 σ 键,并且在同一个平面上,每个碳原子和硫原子都有一个 p 轨道,互相平行,碳原子的 p 轨道中有一个 p 电子,硫原子的 p 轨道中有两个 p 电子,形成了一个环状封闭的 6π 电子共轭体系,符合休克尔规则,具有与苯类似的芳香性。

噻吩可以发生硝化、磺化、卤化、Friedel-Crafts 酰基化、Friedel-Crafts 烷基化等亲电取代反应,以及加成反应。噻吩与强酸在一起,也能够破坏其共轭体系而发生水解等反应。但噻吩比呋喃稳定,不易发生聚合反应。

噻吩在香料中的重要地位是最近 40 年才被人们所认识的,它们在许多食品中已被检测出来,对食品的感官特性有极大的贡献。在罐头牛肉、烹调猪肝、炸仔鸡、烧鸡肉、炒洋葱、炒花生、炒榛子、烤面包、爆玉米、熟米饭、焙烤咖啡、茶叶、啤酒等许多动植物食品中,均有微量噻吩及噻吩酮的存在。2-乙酰基-3-甲基噻吩可以对糖浆提供蜜样香味。3-乙酰基-2-甲基噻吩可用来改进食品和香烟的香气和气味。双噻吩基硫化物可用于制备坚果型香精,用于改进花生、核桃、肉、肉汤及蘑菇的香味。2,5-二甲基-4-羟基-3(2H)-噻吩酮可以作为烤肉和焦肉香味的增香剂。

本章内容涉及含有噻吩环、二氢噻吩环、四氢噻吩环的化合物。

第一节　噻吩类化合物的一般合成方法

一、由呋喃类化合物合成噻吩类化合物

含呋喃环的化合物与一些含硫试剂反应可以转化为相应的含噻吩环的化合物。例如[1,2]:

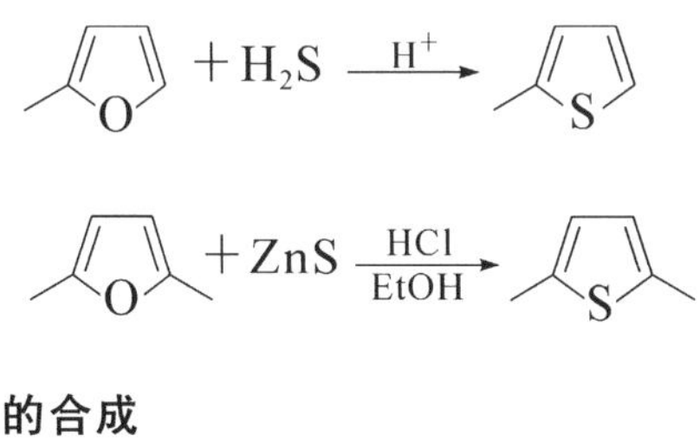

二、酰基噻吩类化合物的合成

乙酰基和多碳酰基噻吩类化合物一般可以用 Friedel-Crafts 反应合成。

$$\text{methylthiophene} + (CH_3CO)_2O \longrightarrow \text{2-acetyl-5-methylthiophene}$$

常用的催化剂有四氯化锡、三氟化硼-乙醚[3,4]、氯化锌[5]等。

甲酰基噻吩类化合物一般可以通过 Vilsmeier 反应合成[6]。

$$\text{thiophene} + Ph-N(CH_3)-CHO \xrightarrow{POCl_3} \text{2-thiophene-CHO} + Ph-NH-CH_3$$

三、2-噻吩酸酯类化合物的合成

2-噻吩酸酯类化合物可以通过 2-噻吩甲酸与醇直接酯化法制备。

$$\text{thiophene-2-COOH} + ROH \xrightarrow[\triangle]{H_2SO_4} \text{thiophene-2-COOR} + H_2O$$

式中：R 为甲基、乙基、正丙基、异丙基、正丁基、异丁基、仲丁基、正戊基、苄基、糠基等[7,8]。

四、3-四氢噻吩酮类化合物的合成

3-四氢噻吩酮类化合物可以通过相应的甲酯基-3-四氢噻吩酮水解、脱羧制备[9]。

$$\text{methyl 3-oxotetrahydrothiophene-2-carboxylate} + H_2O \xrightarrow[100℃]{H_2SO_4} \text{3-oxotetrahydrothiophene} + CO_2 + CH_3OH$$

五、2-巯基噻吩衍生物的合成

2-巯基噻吩本身可用作食品香料，以 2-巯基噻吩为原料，可以合成一系列噻吩类香料。例如，2-巯基噻吩氧化可以合成二(2-噻吩基)二硫醚[10]。

$$\text{thiophene-2-SH} + I_2 \xrightarrow[Na_2CO_3]{KI} \text{thiophene-S-S-thiophene}$$

2-巯基噻吩与二氯化二硫反应可以制取二(2-噻吩基)四硫醚[10]。

$$\text{thiophene-2-SH} + S_2Cl_2 \longrightarrow \text{thiophene-S-S-S-S-thiophene}$$

六、3-巯基二氢噻吩类化合物的合成

3-巯基二氢噻吩类化合物一般具有肉香和烤肉香,可以通过相应的四氢噻吩-3-酮制备[11]。

$$\text{四氢噻吩-3-酮} + H_2S \xrightarrow[-80℃]{CH_3CH_2OH、HCl} \text{3-巯基噻吩衍生物}$$

$$\text{2-甲基四氢噻吩-3-酮} + H_2S \xrightarrow[-80℃]{CH_3CH_2OH、HCl} \text{2-甲基-3-巯基噻吩衍生物}$$

七、2-(2-噻吩基)乙硫醇及其乙酸酯的合成

2-(2-噻吩基)乙硫醇及其乙酸酯都具有焦香和洋葱香味,可以通过 2-乙烯基噻吩制备[12]。

$$\text{2-乙烯基噻吩} + HS-C(=O)CH_3 \longrightarrow \text{噻吩-CH}_2\text{CH}_2\text{-S-C(=O)CH}_3 \xrightarrow{H_2O} \text{噻吩-CH}_2\text{CH}_2\text{SH}$$

八、2-噻吩甲基醚类化合物的合成

2-噻吩甲基醚类化合物一般可以通过 2-氯甲基噻吩与相应的醇反应制备。

$$\text{噻吩-CH}_2\text{Cl} + ROH \xrightarrow{KOH} \text{噻吩-CH}_2\text{-O-R}$$

式中:R 为糠基或 2-噻吩亚甲基[13]。

九、2-噻吩甲醇酯类化合物的合成

2-噻吩甲醇酯类化合物可以由 2-噻吩甲醇与相应的酸酐反应制得[13]。

$$\text{噻吩-CH}_2OH + (CH_3CO)_2O \longrightarrow \text{噻吩-CH}_2\text{-O-C(=O)CH}_3 + CH_3COOH$$

十、3-烷氧基噻吩类化合物的合成

3-甲氧基噻吩可以通过 3-溴噻吩与甲醇钠反应合成。3-烷氧基噻吩类化合物可以通过 3-甲氧基噻吩与醇发生取代反应合成。

$$\text{3-溴噻吩} + CH_3ONa \longrightarrow \text{3-甲氧基噻吩} \xrightarrow{ROH} \text{3-烷氧基噻吩}$$

式中:R 为异丁基、异戊基、异辛基等[14]。

第二节 噻吩类香料

一、5-甲基-2-噻吩基甲醛

5-甲基-2-噻吩基甲醛(5-methyl-2-thiophenecarboxaldehyde, FEMA 号 3209)天然存在于炸土豆片、炒花生、西红柿、面包、生鸡肉、威士忌、咖啡等食品中,具有甜的、杏仁、樱桃、木香香气,杏仁、樱桃、坚果味道,常用于浆果、杏仁、榛子、黑莓、樱桃、香草等食品香精,在加香食品中的建议用量为 0.5mg/kg[15]。

5-甲基-2-噻吩基甲醛可以用 2-甲基噻吩为原料,通过 Vilsmeier 反应合成[6]。

$$\text{噻吩-CH}_3 + \text{Ph-N(CH}_3\text{)-CHO} \xrightarrow{POCl_3} \text{噻吩(CH}_3\text{)-CHO} + \text{Ph-NH-CH}_3$$

二、2-噻吩硫醇

2-噻吩硫醇(thienyl mercaptan, FEMA 号 3062)具有焦糊橡胶、烤咖啡香气以及焦糊或烤香味道[15],常用于咖啡、烤香等食品香精,在加香食品中浓度约为 0.1mg/kg[16]。

2-噻吩硫醇可以由 2-噻吩磺酰氯还原制备或 2-氯噻吩与硫化氢反应制备[17]。

$$\text{噻吩-Cl} + H_2S \longrightarrow \text{噻吩-SH} + HCl$$

噻吩类香料化合物大多具有焦香、肉香、坚果香、葱蒜香。表 10-1 列出部分噻吩类香料的化学结构和香味特征。

表 10-1 部分噻吩类香料的化学结构和香味特征

名称	化学结构式	香味特征或应用建议
2-甲基噻吩		烤香、青香、洋葱香;苦味[13,18]
2-甲基-4-乙基噻吩		焦香[13]
2-甲基-5-丙基噻吩		类似洋葱[13]
3-甲基噻吩		脂肪样、葡萄样[13]

第十章 噻吩类香料

续表

名称	化学结构式	香味特征或应用建议
2-甲基-2,3(或 2,5)-二氢噻吩		甘蓝样[18]
2,5-二甲基噻吩		青香[13]
3,4-二甲基噻吩		洋葱香
2-乙基噻吩		类似苏合香烯[13]
3-乙基噻吩		类似苏合香烯[13]
2-乙烯基噻吩		类似苏合香烯[13]
玫瑰噻吩		柑橘香[19]
2-噻吩甲醛		苯甲醛样味道[13]
3-甲基-2-噻吩甲醛		樟脑香韵,藏红花味道[13]
5-甲基-2-噻吩甲醛（FEMA 号 3209）		杏仁香,樱桃样[20]。可用于饮料、糖果、焙烤、肉制品、奶制品等香精[21]
5-丙基-2-噻吩甲醛		焦糖香味[13]
5-(2-亚甲基噻吩)-2-噻吩甲醛		焦糊香,焦糖香味[13]
2-乙酰基噻吩		炒麦芽香,类似洋葱香味[20]
5-甲基-2-乙酰基噻吩		甜的,花香[20]
3-甲基-2-乙酰基噻吩		坚果香
2,5-二甲基-3-乙酰基噻吩（FEMA 号 3527）		可用于焙烤食品、奶制品、肉制品、饮料、糖果等香精[22]

续表

名称	化学结构式	香味特征或应用建议
四氢噻吩-3-酮 (FEMA 号 3266)		葱香、烤香、烧肉香、蔬菜香气；洋葱、大蒜、烧肉、蔬菜味道[20]。可用于奶制品、肉制品、饮料、糖果等香精[23]
2-甲基四氢噻吩-3-酮 (FEMA 号 3512)		青香、芳香、焦香、咖啡香味[20]
4-羟基-5-甲基-3(2H)-噻吩酮		烤肉香、爆玉米香[18]
4-巯基-5-甲基-3(2H)-噻吩酮		坚果香[18]
5-甲基-2-(5-甲基-2-噻吩基)噻吩		果香、青香[13]
二(2-噻吩基)甲烷		果香、青香[13]
二(5-甲基-2-噻吩基)甲烷		果香[13]
2-噻吩基-2-呋喃基甲烷		青香[13]
5-甲基-2-噻吩基 2-呋喃基甲烷		干果香[13]
二(2-噻吩基)二硫醚 (FEMA 号 3323)		坚果香、肉香、壤香、烤肉香味[10]
二(2-噻吩基)四硫醚		肉汤香味[24]
甲基 2-噻吩甲基醚		芥末样味道[13]
二(2-噻吩甲基)醚		果香、木香[8,13]
糠基 2-噻吩甲基醚		木香、青香[13]
2-噻吩酸丙酯		焦香味[8]

第十章 噻吩类香料

续表

名称	化学结构式	香味特征或应用建议
2-噻吩酸糠酯	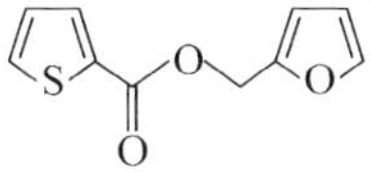	壤香[8]
2-硫代噻吩酸甲酯	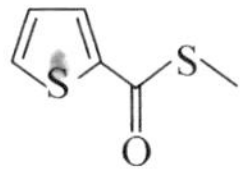	煮蔬菜样味道[8]
2-硫代噻吩酸乙酯	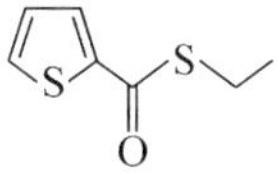	焦糊,咖啡样味道[8]
2-硫代噻吩酸糠酯	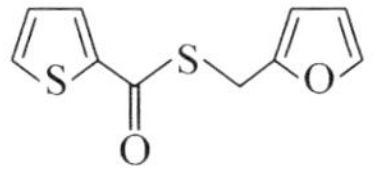	咖啡样[8]
3-甲氧基噻吩	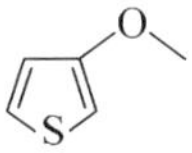	硫磺味[14]
3-(2-甲基丙氧基)噻吩	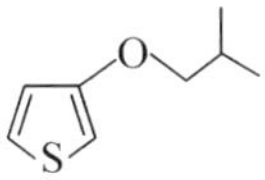	大蒜味、洋葱味[14]
3-(3-甲基丁氧基)噻吩	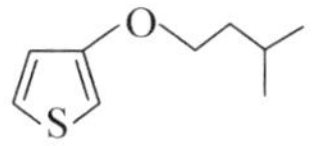	烤肉香[14]
3-(2-乙基己氧基)噻吩	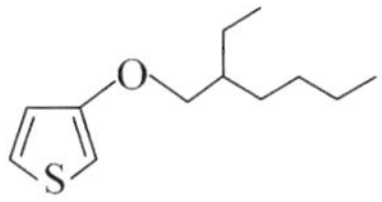	烤肉香[14]
3-戊氧基噻吩	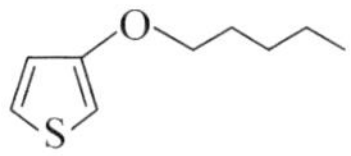	烤肉香味[25]
3-己氧基噻吩	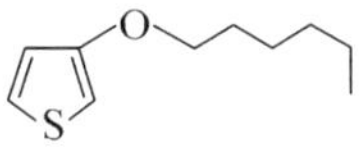	烤肉香味[25]
3-辛氧基噻吩	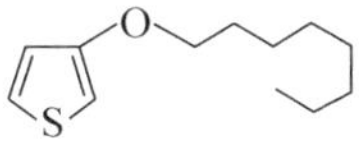	烤肉香味[25]
3-噻吩硫醇	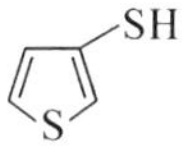	脂香、洋葱香、熟肉风味特征,似咖啡香[26,27]
2-甲基-3-巯基噻吩	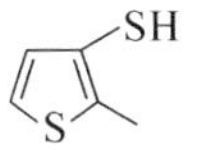	烤肉香、烤香、熟肉特征香、似蛋香、脂香、似洋葱香、似韭葱香[18,26,28]
2-甲基-4-巯基噻吩	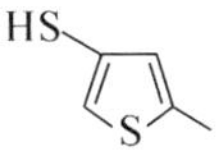	橡胶样[18,28]
2-(2-噻吩基)乙硫醇	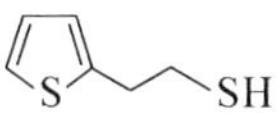	焦香、咖啡香、洋葱香味[12]

续表

名称	化学结构式	香味特征或应用建议
硫代乙酸 2-噻吩乙酯		焦香,洋葱香味[12]
2-甲基-3-巯基-2,3-二氢噻吩		甜的,烤肉香[18,28]
2-甲基-3-巯基-4,5-二氢噻吩		肉香[18,28]
2-甲基-4-巯基-2,3-二氢噻吩		橡胶样、肉香[18,28]
2-甲基-4-巯基-4,5-二氢噻吩		烤肉香[18,28]
2-甲基-3-巯基-4-羟基-2,3-二氢噻吩		肉香、肉汁香[18,28]
2-甲基-4-巯基四氢噻吩		肉样[18,28]
顺-2-甲基-3-巯基四氢噻吩		肉香[18,28]
反-2-甲基-3-巯基四氢噻吩		肉香、肉汁香[18,28]
2-甲基-3-呋喃基 2-甲基-3-噻吩基二硫醚		肉香、洋葱香、大蒜香、金属样、脂肪香[28,29]
二(2-甲基-3-噻吩基)二硫醚		硫化物样、金属样、橡胶样、淡肉香[28]
甲基 2-甲基-3-噻吩基三硫醚		肉香[30]
2-甲基-3-呋喃基 顺-2-甲基-3-四氢噻吩基硫醚		坚果香、蘑菇香、肉香[31]
2-甲基-3-呋喃基 反-2-甲基-3-四氢噻吩基硫醚		烤香、蔬菜香、蘑菇香、肉香[31]

续表

名称	化学结构式	香味特征或应用建议
2-甲基-3-噻吩基 顺-2-甲基-3-四氢噻吩基硫醚		蘑菇香、肉香、热带水果香[31]
2-甲基-3-噻吩基 反-2-甲基-3-四氢噻吩基硫醚		烤肉香、焦糊的肉香、肉香[31]
2-甲基-3-呋喃基 2-亚甲基四氢噻吩基硫醚		大蒜香、洋葱香、淡肉香[31]
2-甲基-3-噻吩基 2-亚甲基四氢噻吩基硫醚		洋葱香、热带水果香、淡肉香[31]
2-甲基-3-噻吩基 2-甲基-2-四氢噻吩基硫醚		洋葱香、甘蓝香、焦香、热带水果香、烤肉特征香气[31]
2-甲基-3-呋喃基 2-甲基-2-四氢噻吩基硫醚		青香、烤香、典型的肉香[31]
2,3,3′,2′-噻吩并噻吩		类似荞麦香、坚果香[13]
苯并噻吩		壤香、橡胶样[13]
2-苯并噻吩甲醛		杏仁香、焦糖香、奶油香、坚果香味[13]
2-乙酰基苯并噻吩		壤香、烤香、甜的香味[13]

参 考 文 献

[1] Cubina T I, Rogacheva S M, Kharchenko V G. New approach to synthesis of 2-methyl-thiophene. Chem. Heterocycl. Compd. (Engl. Transl.), 1999, 35(2): 244

[2] Cubina T I, Drevko B I, Fedina L N et al. New method for synthesis of 2,5-disubstituted thiophenes. Chem. Heterocycl. Compd. (Engl. Transl.), 1999, 35(6): 650~652

[3] Farrar M W, Levine R. Condensations effected by boron fluoride complexes. Ⅲ. The Acylation of certain substituted thiophenes and furans. J. Am. Chem. Soc., 1950, 72: 3695~3698

[4] Levin R, Heid J V, Farrar M W. Further studies in the acylation of thiophene and furan in

the presence of boron fluoride complexes. J. Am. Chem. Soc., 1949, 71: 1207~1209

[5] Hartough H D, Kosak A I. Acylation studies in the thiophene and furan series. J. Am. Chem. Soc., 1947, 69: 1012~1013

[6] Loudon G M. Conversion of aliphatic amides into amines with [1,1-bis(trifluoroacetoxy) iodo]benzene. Organic Chemistry, 1984, 49(22): 4272~4276

[7] Weinstein B. Some esters of 2-thenoic acid. J. Am. Chem. Soc., 1955, 77: 6709

[8] Winter M, Gautschi F, Flament I et al. Flavor modified soluble coffee: US, 3702253. 1972.11.7

[9] Woodward R B, Eastman R H. Tetrahydrothiophene derivatives. J. Am. Chem. Soc., 1946, 68: 2229~2235

[10] Katz I, Evers W J, Giacino C. Sulfur-containing flavoring compositions and processes therefore: US, 3706577. 1972.12.19

[11] Ouweland V D, Peer H G. Flavoring substances: US, 4080367. 1978.3.21

[12] Torrey S. Fragrances and Flavors, Recent Developments. Park Ridge, New Jerssy: Noyes Data Corporation, 1980. 299~301

[13] Winter M, Goldman I M, Gautschi F et al. Flavoring agent: US, 3943260. 1976.3.9

[14] 黄金波,余爱农,孙宝国等. 3-异烷氧基噻吩的合成. 精细化工, 2006, 23(3): 282~283

[15] 孙宝国,何坚. 香精概论. 第二版. 北京: 化学工业出版社, 2006. 112~121

[16] Hall R L, Oser B L. Recent progress in the consideration of flavoring ingredients under the food additives amendment. 3. GRAS Substances. Food Technology, 1965, 19(2): 151~197

[17] 孙宝国,何坚. 香料化学与工艺学. 第二版. 北京:化学工业出版社, 2004. 409~410

[18] Godefridus A M, Ouweland V D, Peer H G. Components contributing to beef flavor. Volatile compounds produced by the reaction of 4-hydroxy-5-methyl-3(2H)-furanone and its thio analog with hydrogen sulfide. J. Agric. Food Chem., 1975, 23(3): 501~505

[19] 宫本正義. 含硫化合物. 香料(日), 1987, (6): 137~139

[20] Winter M, Flament I, Stoll M et al. Flavor modified soluble coffee: US, 3702253. 1972.11.7

[21] Hall R L, Oser B L. Recent progress in the consideration of flavoring ingredients under the food additives amendment. 4. GRAS Substances. Food Technology, 1970, 24(5): 25~34

[22] Oser B L, Ford R A. Recent progress in the consideration of flavoring ingredients under the food additives amendment. 11. GRAS substances. Food Technology, 1978, 32(2): 60~70

[23] Oser B L, Hall R L. Recent progress in the consideration of flavoring ingredients under the food additives amendment. 5. GRAS Substances. Food Technology. 1972, 26(5): 36~41

[24] 黄荣初,王兴凤. 肉类香味的合成香料. 有机化学,1983,(3):175~179

[25] 黄金波,余爱农,孙宝国等. 无水硫酸氢钾催化合成3-烷氧基噻吩. 化学试剂,2006,28(4):219~221,234
[26] Shahidi F. 肉制品与水产品的风味. 李洁,朱国斌译. 北京:中国轻工业出版社,2001. 66~67
[27] 孙宝国,田红玉,郑福平等. 3-噻吩硫化物及其衍生物分子结构与香味的关系研究//中国香料香精化妆品工业协会. 2004年中国香料香精学术研讨会论文集. 北京:中国香料香精化妆品工业协会,2004. 198~201
[28] Macleod G. The flavor of beef//Shahidi F. Flavor of Meat, Meat Products and Seafoods. London:Blackie Academic & Professional,1998. 28~60
[29] Macleod G. The flavor of beef//Shahidi F. Flavor of Meat and Meat products. New York:Blackie Academic & Professional,1994. 6~7
[30] Güntert M,Bertram H J,Emberger R et al. Thermal degradation of thiamin(Vitamin B_1)//Mussinan C J,Keelan M E. Sulfur Compounds in Foods. Washington DC:American Chemical Society,1994. 220~222
[31] Werkhoff P,Brüning J,Emberger R et al. Isolation and characterization of volatile sulfur-containing meat flavor components in model systems. J. Agric. Food Chem.,1990. 38(3):777~791

第十一章 噻唑类香料

噻唑可以看成是噻吩环上 3-位的 CH 换成 N，这个氮原子用 sp^2 杂化轨道成键，还有一个 p 轨道中有一个 p 电子占据，并参加共轭，形成了一个环状封闭的 6π 电子共轭体系，符合休克尔规则，具有一些芳香性。此外，该氮原子有一个 sp^2 杂化轨道被一对电子占据，未参加成键，具有碱性，pK_b 为 2.4，可以与质子结合。

噻唑不容易发生亲电取代反应，磺化反应需要较强烈的条件；卤化、硝化必须有给电子基团才能发生。

噻唑类化合物广泛存在于天然食品中，1967 年，Stoll 等从可可萃取液中发现了微量的 4-甲基-5-乙烯基噻唑，该化合物具有十分强烈的坚果香气，他们判断噻唑类化合物可以用于香料工业[1]。1972 年，4-甲基-5-乙烯基噻唑获得了 FEMA 号(3313)[2]，成为一种允许使用的食品香料。迄今为止，在炖牛肉、熟牛肝、熟肉干、烤土豆、芦笋、红豆、炒芝麻、芝麻油、炒花生、炒榛子、爆玉米、米饭、牛奶、可可、咖啡、茶叶、葡萄酒、老姆酒、威士忌、啤酒、麦芽、西红柿等食品中都发现微量噻唑类化合物的存在[3]。例如，在芝麻油的挥发性香成分中发现了噻唑、2-甲基噻唑、4-甲基噻唑、5-甲基噻唑、2,4-二甲基噻唑、2,5-二甲基噻唑、4,5-二甲基噻唑、4-甲基-2-乙基噻唑、2,4-二甲基-5-乙基噻唑、2-乙酰基噻唑、4-甲基-2-乙酰基噻唑、5-甲基-2-乙酰基噻唑、4-乙酰基噻唑、苯并噻唑 14 种噻唑类化合物[4]。

维生素 B_1(盐酸硫胺素，FEMA 号 3322)分子中含有噻唑环结构单元，一些重要噻唑类香料化合物和其他重要含硫香料化合物可以在维生素 B_1 热降解过程中生成，如 4-甲基-5-羟乙基噻唑的生成过程如图 11-1 所示[5]。

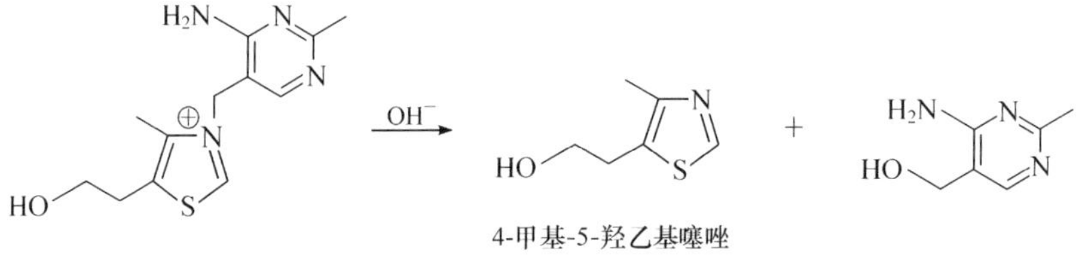

图 11-1 由维生素 B_1 生成 4-甲基-5-羟乙基噻唑

噻唑类化合物一般具有新鲜蔬菜、西红柿、烤肉、坚果、爆玉米花等香味特征，在咸味香精中占有重要地位，在对中国早期牛肉香精的挥发性成分剖析中就发现了 2-乙酰基噻唑和 4-甲基-5-羟乙基噻唑等重要噻唑类香料化合物[6]，这些噻唑类香料现在仍然是咸味香精的重要香料。

第一节 噻唑类化合物的一般合成方法

一、α-卤代酮和硫代酰胺反应

α-卤代酮和硫代酰胺反应可以合成多种取代的噻唑类化合物[7]。例如：

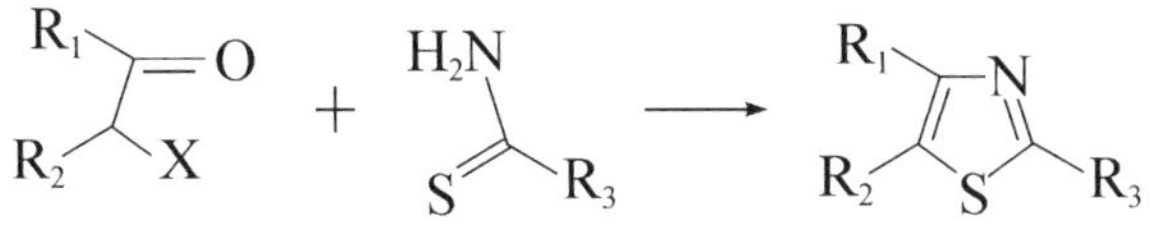

二、α-卤代酮与酰胺和五硫化二磷反应

α-卤代酮与酰胺和五硫化二磷反应也可以合成多种取代的噻唑类化合物。例如：

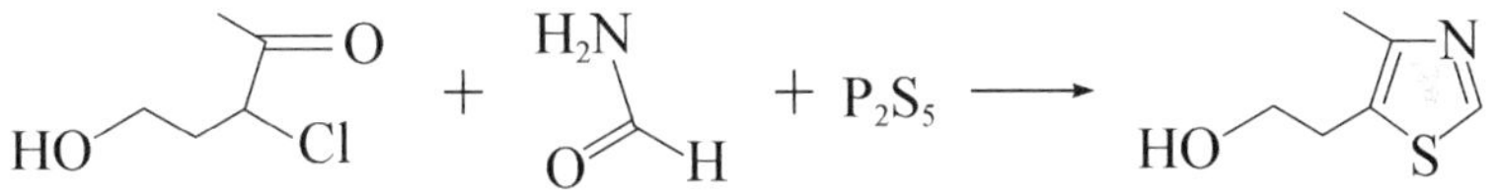

三、由 α-酰胺基酮合成

α-酰胺基酮或 α-酰胺基酯等在五硫化二磷存在下环化得到 2,4,5-三烷基噻唑类化合物或 2,4-二烷基-5-烷氧基噻唑类化合物[8]。

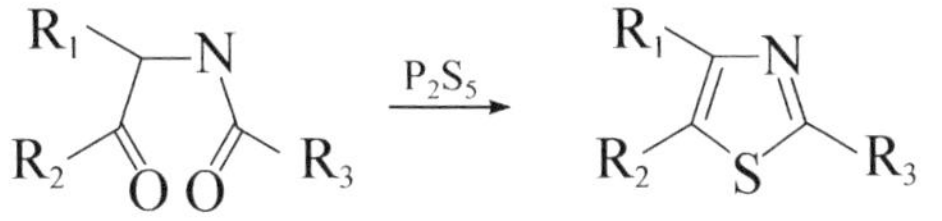

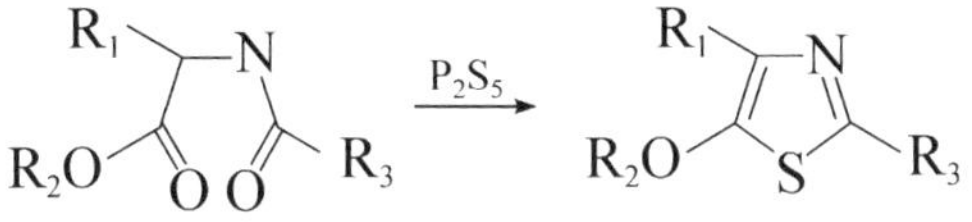

四、由 2,5-二羟基-1,4-二噻烷合成

2,5-二羟基-1,4-二噻烷与氨和醛反应先生成 2-烷基-2,5-二氢噻唑，后脱氢得到 2-烷基噻唑[9]。

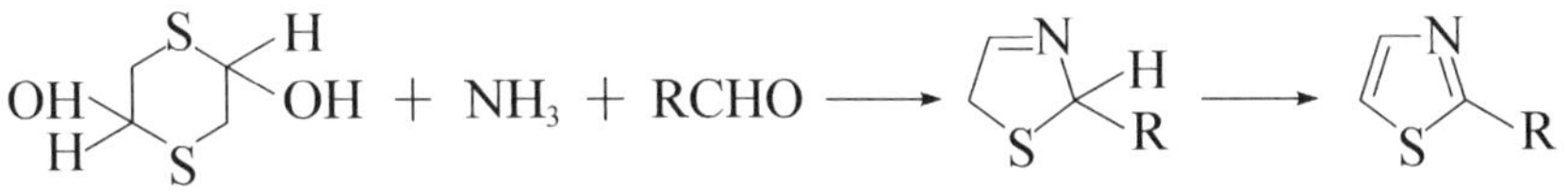

五、由 α-巯基醛(酮)与腈反应合成

α-巯基醛(酮)与腈在干燥的 HCl 条件下反应得到噻唑[10]。

式中：R_1，R_2，R_3 为烷基或 H。

六、由 α-巯基醛（酮）与 NH_3 和醛反应合成

α-巯基醛（酮）在 NH_3 存在下与醛反应生成噻唑啉，然后脱氢得到噻唑[11]。

七、由三甲基-2-噻唑基硅烷合成

以三甲基-2-噻唑基硅烷（2-trimethylsilylthiazole）为原料，可以合成多种噻唑类化合物[12,13]，如三甲基 2-噻唑基硅烷与烷基酰氯或醛反应，分别生成 2-噻唑基烷基酮和 2-噻唑基烷基甲醇。

八、由 4-甲基噻唑合成 4-甲基-2-噻唑基烃基甲醇

4-甲基噻唑先与丁基锂反应，生成物再与醛反应可以合成 4-甲基-2-噻唑基烃基甲醇类化合物。例如：

式中：R 为甲基、乙基、乙烯基、异丁基等[14]。

九、由 4-甲基噻唑合成 4-甲基-2-酰基噻唑

4-甲基噻唑先与溴化乙基镁反应，生成的格氏中间体再与酰卤或酸酐反应得到

4-甲基-2-烷酰基噻唑类化合物[14]。例如,与丙酰氯反应生成 4-甲基-2-丙酰基噻唑。

$$\underset{S}{\overset{N}{\diagdown}}\text{-CH}_3 + C_2H_5MgBr + \underset{O}{\overset{Cl}{\diagdown}}\text{-C}_2H_5 \longrightarrow \underset{S}{\overset{N}{\diagdown}}\text{-CO-C}_2H_5$$

十、由 2-卤代噻唑合成

α-硫氰基酮在卤代酸存在下环化生成 2-卤代噻唑,它是制备各种噻唑衍生物的中间体。

$$\underset{R_2}{\overset{R_1}{\diagdown}}\text{C=N} \xrightarrow{HX} \underset{R_2}{\overset{R_1}{\diagdown}}\text{-X}$$

2-卤代噻唑用锌和乙酸还原,可以得到 2-位没有取代的噻唑。

$$\underset{R_2}{\overset{R_1}{\diagdown}}\text{-X} \xrightarrow{Zn+CH_3COOH} \underset{R_2}{\overset{R_1}{\diagdown}}\text{-H}$$

2-卤代噻唑与硫脲反应,可以得到 2-巯基噻唑。2-巯基噻唑与卤代烷反应得到 2-烷硫基噻唑[15]。2-巯基噻唑用碘氧化得到相应的二硫醚[16]。例如,4-甲基-2-巯基噻唑氧化得到二(4-甲基-2-噻唑基)二硫醚。

$$\underset{R_2}{\overset{R_1}{\diagdown}}\text{-X} \xrightarrow{(NH_2)_2CS} \underset{R_2}{\overset{R_1}{\diagdown}}\text{-SH} \xrightarrow{R_3X} \underset{R_2}{\overset{R_1}{\diagdown}}\text{-SR}_3$$

$$\underset{S}{\overset{N}{\diagdown}}\text{-SH} \xrightarrow{[O]} \underset{S}{\overset{N}{\diagdown}}\text{-SS-}\underset{S}{\overset{N}{\diagdown}}$$

2-卤代噻唑与烷基硫化钠反应,生成相应的 2-烷硫基噻唑。

$$\underset{R_2}{\overset{R_1}{\diagdown}}\text{-X} + R_3SNa \longrightarrow \underset{R_2}{\overset{R_1}{\diagdown}}\text{-SR}_3$$

2-卤代噻唑与醇钠反应,可以生成相应的 2-烷氧基噻唑。

$$\underset{R_2}{\overset{R_1}{\diagdown}}\text{-X} + R_3ONa \longrightarrow \underset{R_2}{\overset{R_1}{\diagdown}}\text{-OR}_3$$

2-卤代噻唑在正丁基锂存在下,用乙醚作溶剂,在-80℃时与乙醛反应,然后再经过氧化,可以合成 2-乙酰基噻唑。

$$\underset{R_2}{\overset{R_1}{\diagup}}\underset{S}{\overset{N}{\diagdown}}X \xrightarrow[C_4H_9Li]{CH_3CHO} \xrightarrow{[O]} \underset{R_2}{\overset{R_1}{\diagup}}\underset{S}{\overset{N}{\diagdown}}\overset{O}{\underset{}{C-CH_3}}$$

十一、由二乙胺或二异丙基胺与硫高温反应制取

由二乙胺或二异丙基胺与硫在 500℃反应,可以制取 2-甲基噻唑或 2,4-二甲基噻唑。

$$Et_2NH + S \xrightarrow{500℃} \text{2-methylthiazole}$$

第二节 烷基噻唑类香料

一、4-甲基噻唑

4-甲基噻唑(4-methylthiazole,FEMA 号 3716)为无色至淡黄色液体,天然存在于煮过的芦笋中,具有葱、蔬菜、西红柿、青香、坚果、热带水果香气以及青香、蔬菜、葱、肉香味道,可用于西红柿、坚果、肉、热带水果等食品香精,在加香食品中的使用浓度一般为 1.5~5.0mg/kg[17]。

4-甲基噻唑可以由氯丙酮与硫代甲酰胺反应制备。

二、4,5-二甲基噻唑

4,5-二甲基噻唑(4,5-dimethylthiazole,FEMA 号 3274)为无色或黄色液体,天然存在于桃子、猪肉、咖啡、虾、扇贝、炒花生中,具有鱼香、菜肴、胺样香气,海鲜味道,可用于虾类、贝类、蟹类、蔬菜、咖啡、坚果及青香等食品香精,在加香食品中的使用浓度一般为 6~20mg/kg[2]。

4,5-二甲基噻唑可以由 3-溴-2-丁酮和硫代甲酰胺制备。

三、2,4,5-三甲基噻唑

2,4,5-三甲基噻唑(2,4,5-trimethylthiazole,FEMA 号 3325)为无色液体,天然存在于牛肉、红豆、煮土豆和葡萄酒中,具有烤香、肉香、坚果、可可、蔬菜香气和味道,主要用于烟草、烤香、坚果、巧克力、可可、咖啡、土豆、蔬菜、肉味等食用香精,在加香食品中的使用浓度一般为 2~6mg/kg[2]。

2,4,5-三甲基噻唑可以由 3-溴-2-丁酮与乙酰胺和 P_2S_5 反应制备[18,19]。

2,4,5-三甲基噻唑也可以由 3-溴-2-丁酮或 3-氯-2-丁酮与硫代乙酰胺反应制备[19]。

表 11-1 列出部分其他烷基噻唑类香料的化学结构和香味特征。

表 11-1 部分其他烷基噻唑类香料的化学结构和香味特征

名称	化学结构式	香味特征
噻唑 (FEMA 号 3615)		似吡啶样香气[20]
2-甲基噻唑		青菜香[20]
2,4-二甲基噻唑		肉香、可可样香气[20]
2-乙基噻唑		青香、坚果香[21]
2-乙基-4-甲基噻唑 (FEMA 号 3680)		坚果香、烤香、咖啡香、菜肴香、肉香香气;咖啡、可可、肉香味道
5-乙基-4-甲基噻唑		坚果香、青香、壤香[21,22]
5-乙基-2,4-二甲基噻唑		坚果香、烤香、肉香[21,22]
2,5-二乙基-4-甲基噻唑		青香、坚果香[21,22]

名称	化学结构式	香味特征
2-异丙基噻唑		青香、蔬菜香[21]
2-异丙基-4-甲基噻唑（FEMA 号 3555）		霉味、壤香、淡果香、咖啡及肉香[23]
2-丙基噻唑		青香、药草、坚果香[21]
2-异丁基噻唑（FEMA 号 3134）		青香、蔬菜香、草香、西红柿、藤叶香气；青香、蔬菜、西红柿味道[22]
4-异丁基-5-乙基噻唑		黄瓜香、青香、似马铃薯香气[20]
2-丁基噻唑		青的药草香[20]
2-(2-丁基)噻唑（FEMA 号 3372）		青香、药草香[21,22]
4-丁基-5-丙基噻唑		甜椒香气[24]

第三节　烷氧基噻唑类香料

一、2-乙氧基噻唑

2-乙氧基噻唑(2-ethoxylthiazole，FEMA 号 3340)为无色透明油状液体，具有霉香、蔬菜、青香、酚香、坚果、咖啡香气和味道，主要用于咖啡、肉味、可可、胡桃、蔬菜等食品香精，在加香食品中的使用浓度为 0.2～2mg/kg[25]。

2-乙氧基噻唑可以由 2-氯代噻唑或 2-溴代噻唑与乙醇钠，在无水乙醇溶剂中加热回流制取[18]。

二、其他烷氧基噻唑类香料

表 11-2 列出部分其他烷氧基噻唑类香料的化学结构和香味特征。

表 11-2 部分其他烷氧基噻唑类香料的化学结构和香味特征

名称	化学结构式	香味特征
2-甲氧基噻唑		甜的、烤香、酚样香气[20,22]
5-甲氧基噻唑		烤香、肉香、洋葱香[22]
2-甲基-5-甲氧基噻唑（FEMA 号 3192）		甘蓝香、硫化物样、蔬菜样香[22]
4-异丁基-5-甲氧基噻唑		蔬菜香、洋葱香、芹菜香、青椒香[18,21,22]
2-甲基-4-异丁基-5-甲氧基噻唑		青香、洋葱香、蔬菜香[18,20,22]
5-乙氧基噻唑		煮洋葱香[22]
4-异丁基-5-乙氧基噻唑		强烈的黄瓜香、青椒香、洋葱香、壤香、青土豆香[18,22]
2-甲基-4-异丁基-5-乙氧基噻唑		青香、蔬菜香、洋葱香气[20,22]
2-丁氧基噻唑		青菜香[20,22]

第四节 酰基噻唑类香料

一、2-乙酰基噻唑

2-乙酰基噻唑（2-acetylthiazole，FEMA 号 3328）为油状液体，天然存在于牛肉汤、炖牛肉、牛肝、猪肝、白面包、芦笋、土豆、米饭中，具有爆玉米、炒板栗、烤麦片香、烤肉、坚果香、面包香味，可用于坚果、麦片、面包、爆玉米、肉味等食品香精，在加香食品中的使用浓度一般为 0.2~1.4mg/kg[25]。

2-乙酰基噻唑可以由 2-溴代噻唑与乙醛在丁基锂存在下经缩合和氧化制取；也可以由三甲基 2-噻唑基硅烷与乙酰氯反应制备[12,13]。

二、2,4-二甲基-5-乙酰基噻唑

2,4-二甲基-5-乙酰基噻唑(2,4-dimethyl-5-acetylthiazole,FEMA 号 3267)为无色透明液体,具有霉香、壤香、咖啡、肉香、坚果香气,烤香、坚果、肉香、菜肴、木香、坚果味道,主要用于咖啡、榛子、坚果、调味品、谷物、巧克力等食品香精,在加香食品中的使用浓度一般为 6~10mg/kg[2]。

2,4-二甲基-5-乙酰基噻唑可以由 3-氯-2,4-戊二酮与硫代乙酰胺反应制备。

表 11-3 列出部分其他酰基噻唑类香料的化学结构和香味特征。

表 11-3　部分其他酰基噻唑类香料的化学结构和香味特征

名称	化学结构式	香味特征或应用建议
4-甲基-2-乙酰基噻唑		焦香味[24]
5-甲基-2-乙酰基噻唑		烤咖啡香[24]
4-乙酰基噻唑		坚果香、谷物香[21,22]
2,5-二甲基-4-乙酰基噻唑		烤香、肉香、硫化物样[21,22]
5-乙酰基噻唑		烤肉香、洋葱香[20]
4-甲基-5-乙酰基噻唑		烤香、坚果香、硫化物样[21,22]

名称	化学结构式	香味特征或应用建议
2,4-二甲基-5-乙酰基噻唑（FEMA 号 3267）		烤香、坚果香、肉香[18,21,22]
2-丙酰基噻唑（FEMA 号 3611）		可用于饮料、糖果、焙烤食品等香精[26]

第五节　二氢噻唑类香料

二氢噻唑又称噻唑啉（thiazoline），香料中常见的是 4,5-二氢噻唑（2-噻唑啉）和 2,5-二氢噻唑（3-噻唑啉）。表 11-4 列出部分二氢噻唑类香料的化学结构和香味特征。

表 11-4　部分二氢噻唑类香料的化学结构和香味特征

名称	化学结构式	香味特征
2-甲硫基-4,5-二氢噻唑		蘑菇香、青香、醚香、壤香[27]
2-乙硫基-4,5-二氢噻唑		果香、烤香[27]
2-正丙硫基-4,5-二氢噻唑		青香、草莓香、壤香、木犀香[27]
2-异丙硫基-4,5-二氢噻唑		椰子香、酸涩、泥土青香、淡果香[27]
2-正丁硫基-4,5-二氢噻唑		茉莉酮香、青香、黑加仑子香、叶香[27]
2-异丁硫基-4,5-二氢噻唑		青香、壤香[27]
2-正戊硫基-4,5-二氢噻唑		壤香、果香、甜烧酒香、烂铁气味、芸香样[27]

续表

名称	化学结构式	香味特征
2-异戊硫基-4,5-二氢噻唑		弱青香、生橡胶样、蒲公英香、烂铁气味、龙蒿香、大茴香香[27]
2-正庚硫基-4,5-二氢噻唑		花香、辛香、茉莉酮香、青香、风信子香、烂铁气味、蒲公英香[27]
2-乙酰基-4,5-二氢噻唑 (FEMA 号 3817)		面包香[22]、麦片香、洋葱香、鸡皮香、坚果香、辛香、草香香气;土豆、爆玉米花、麦片味道[28]
2-丙酰基-4,5-二氢噻唑 (FEMA 号 4064)		肉香、烤香、爆玉米花香气[23]
2,4,5-三甲基-2,5-二氢噻唑		肉香、坚果香、洋葱气息[29]
4,5-二甲基-2-乙基-2,5-二氢噻唑 (FEMA 号 3620)		咖啡香、焦香、巧克力香、吡嗪样、生蔬菜香气[28]
4,5-二甲基-2-异丁基-2,5-二氢噻唑 (FEMA 号 3621)		坚果香、烤香香气;坚果、烤香、肉香、焦糊的谷物味道[28]
4,5-二甲基-2-(2-丁基)-2,5-二氢噻唑 (FEMA 号 3619)		肉香、辛香、蔬菜香[28]
2,4-二甲基-2-巯基甲基-2,5-二氢噻唑		炸洋葱香、肉香[24]

第六节 其他噻唑类香料

一、4-甲基-5-羟乙基噻唑

4-甲基-5-羟乙基噻唑(4-methyl-5-thiazoleethanol,FEMA 号 3204)俗称噻唑醇,为无色或浅黄色黏稠油状液体,具有骨头、骨髓、肉香、坚果、酵母、面包等香气以及肉香、煮肉、烤香、坚果等味道。天然发现于烤牛肉、啤酒中,主要用于坚果、肉类、调味品、面包、乳制品、巧克力、酵母等食品香精,在加香食品中的使用浓度一般为 55mg/kg[30]。

4-甲基-5-羟乙基噻唑可以由硫代甲酰胺与 3-卤-3-乙酰丙醇,或 3-乙酰丙醇,或 γ-丁内酯,或 α-氯代-α-乙酰-γ-丁内酯,或 2-甲基-2,3-二氯-4,5-二氢呋喃等反应制备;也可以由 4-甲基-5-乙酰氧乙基噻唑水解[31]或用 $LiAlH_4$ 还原[32]制备。例如:

二、4-甲基-5-乙酰氧乙基噻唑

4-甲基-5-乙酰氧乙基噻唑（4-methyl-5-thiazoleethanol acetate，FEMA 号 3205）俗称噻唑酯，为无色至黄色黏稠液体，具有甜的、坚果、烤香、豆香、奶香、面包、肉香、肉汤、火腿香气和味道，主要用于火腿、肉味、可可、坚果、面包、奶味等食品香精，在加香食品中的使用浓度一般为 55mg/kg[30]。

4-甲基-5-乙酰氧乙基噻唑可以由 3-氯-4-氧代-1-戊醇乙酸酯与硫代甲酰胺反应制备，或由 4-甲基-5-羟基乙基噻唑与乙酸酐反应制备。

表 11-5 列出部分其他噻唑类香料的化学结构和香味特征。

表 11-5 部分其他噻唑类香料的化学结构和香味特征

名称	化学结构式	香味特征或应用建议
5-羟乙基噻唑		熟洋葱香[20]
2-糠硫基噻唑		青香、蘑菇香、灌木香、樱桃核香味[27]
2-异戊硫基噻唑		生果香、青豌豆香、橙叶油香、甘草香、蜗牛香气[27]
4-甲基-5-乙烯基噻唑（FEMA 号 3313）		焦香、霉臭、坚果香、榛子香、类似蔬菜根、泥土的气味[20]
2,4-二甲基-5-乙烯基噻唑（FEMA 号 3145）		可用于饮料、糖果、焙烤食品等香精[30]

续表

名称	化学结构式	香味特征或应用建议
4-甲基-5-烯丙基噻唑		强烈的坚果样香味[24]
2,4-二甲基-5-烯丙基噻唑 (FEMA 号 3145)		强烈的坚果香味[24]
苯并噻唑 (FEMA 号 3256)		肉香、蔬菜香、咖啡香、坚果香[23]
2-甲硫基苯并噻唑		肉香[24]

参 考 文 献

[1] Stoll M, Dietrich P, Sundt E et al. Recherches sur les arômes. Helv. Chim. Acta, 1967, 50(213~214): 2065~2067

[2] Oser B L, Hall R L. Recent progress in the consideration of flavoring ingredients under the food additives amendment. 5. GRAS substances. Food Technology, 1972, 26(5): 35~42

[3] Buttery R G, Ling L C. Alkylthiazoles in potato products. J. Agr. Food Chem., 1974, 22(5): 912~914

[4] Nakamura S, Nishimura O, Masuda H et al. Identification of volatile flavor components of roasted sesame oil//Bhattacharyya S C, Sen N, Sethi K L. 11th International Congress of Essential Oils, Fragrances and Flavours. Proceedings: Volume 5, Chemistry—Analysis, Structure and Synthesis. New Delhi: Oxford & Ibh publishing Co. Pvt. Ltd., 1989: 73~87

[5] Ashurst P R. Food Flavorings. Second edition. New York: Blackie Academic & Professional, 1995. 148~149

[6] 王迅. 肉味香精挥发性成分剖析及肉香味汤料配制的研究. 1.肉味香精挥发性成分剖析. 食品与发酵工业, 1988, (1): 1~13

[7] Kurkjy R P, Brown E V. The preparation of methylthiazoles. J. Am. Chem. Soc., 1952, 74: 5778~5779

[8] Tarbell D S, Hirschler H P, Carlin R B. 2-Methyl-5-ethoxythiazole and related compounds. J. Am. Chem. Soc., 1950, 72: 3138~3140

[9] Dubs P, Pisaro M. An efficient synthesis of 2-substituted 1,3-thiaziles. Synthesis, 1974, (4): 294~295

[10] 戚建军. 噻唑类香料化合物合成的研究[硕士学位论文]. 北京:北京轻工业学院, 1993. 5~6

[11] Dubs P, Pesaro M. Anefficient synthesis of 2-substituted-1,3-thiazoles. Synthesis, 1974, (4): 294~300
[12] Medici A, Fantin G, Fogagnolo M et al. Reaction of 2-trimethylsilylthiazole with acyl chlorides and aldehydes synt hesis of new thiazole-2-yl derivatives. Tetrahedron Letters, 1983, 24(28): 2901~2904
[13] Medici A, Pedrini P, Dondoni A. Reactions of trimethylsilylthiazole with ketens: a new route to regioselective functionalisation of the thiazoles ring. J. Chem. Soc., Chem. Commun., 1981, 655~656
[14] Max W, Goldman I M. flavoring agent: US, 3943260. 1976.3.9
[15] Stewart F D, Mathes R A. Synthesis of derivatives of 4,5-dimethyl-2-mercaptothiazole. J. Org. Chem., 1949, 14: 1111~1117
[16] Gibbs E M, Robinson F A. Sulphur derivatives of thiazoles. J. Chem. Soc., 1945, 925~928
[17] Oser B L, Ford R A, Bernard B K. Recent progress in the consideration of flavoring ingredients under the food additives amendment. 13. GRAS substances. Food Technology, 1984, 38(10): 66~89
[18] Torrey S. Fragrances and Flavors, Recent developments. Park Ridge. New Jersey: Noyes Data Corporation, 1980. 277~279
[19] 毛多斌,张槐苓,贾春晓等. 三甲基噻唑的合成及在烟用香精中的应用. 郑州轻工业学院学报, 1995, 10(1):76~79
[20] 陈艺易. 调香中应用的噻唑类化合物. 香料香精化妆品, 1992, (2): 66~71
[21] Bedoukian P Z. Perfumery and Flavoring Materials. Wheaton, IL: Allured Publishing Corporation, 1982. 296~310
[22] Pittet A O, Hruza D E. Comparatives study of flavor properties of thiazole derivatives. J. Agr. Food Chem., 1974, 22(2): 264~269
[23] 孙宝国,何坚. 香精概论. 第二版. 北京:化学工业出版社, 2006. 124~153
[24] 黄小凤,李晓东,李中林. 杂环类香料的现状与展望. 化学通报, 1995, (8): 1~16
[25] Oser B L, Ford R A. Recent progress in the consideration of flavoring ingredients under the food additives amendment. 6. GRAS substances. Food Technology, 1973, 27(1): 64~67
[26] Oser B L, Richard A. Ford. Recent progress in the consideration of flavoring ingredients under the food additives amendment, 12. GRAS Substances. Food Technology, 1979, 37(7): 65~73
[27] Vernin G, Zhamkotsian R M, Metzger J. 相转移催化在芳香分子合成方面的应用. 香料与香精, 1983,(4):17~44
[28] 舒宏福. 咸味香料(下). 香料香精化妆品, 2006, (1): 22~26
[29] 金其璋. 食品香料的新成员——杂环化合物. 化学通报, 1984, (1): 31~37
[30] Hall R L, Oser B L. Recent progress in the consideration of flavoring ingredients under

the food additives amendment. 4. GRAS Substances. Food Technology, 1970, 24(5): 25~34
- [31] 陈新志. 4-甲基-5-(β-羟乙基)-噻唑的合成. 浙江化工, 1996, 27(3): 7~9
- [32] Eusebi A J, Brown E V, Cerecedo L R. A new method for the preparation of the thiazole moiety of thiamine. J. Am. Chem. Soc., 1949, 71: 2931~2933

第十二章　其他含硫香料

迄今为止，允许作为食品香料使用的含硫香料化合物都是由 S 和 C、H、O、N 四种元素中的一种或几种组成的，其中仅由 S 和另外一种元素组成的只有 H_2S（FEMA 号 3779）[1]和 SO_2（FEMA 号 3039）[2]，但它们都不是典型的含硫香料。除了前几章介绍的按化学结构比较容易归类的各种含硫香料外，其他含硫香料品种也很多，并且是含硫香料今后研究和发展的热点之一，本章对这些化合物进行简单介绍。

第一节　α,α-二硫醇类香料

1,1-乙二硫醇（FEMA 号 4111）具有硫磺样、橡胶样、洋葱、肉香、烤花生样香气，是猪肉[3]、酵母提出物[4]的挥发性香成分，可用于饮料、肉制品、调味品等香精[5]。二苄基甲二硫醇既可以直接用作香料，也可以作为合成其他杂环香料的原料[6]。该香料可以通过二苄基甲酮与硫化氢反应制备。

第二节　其他结构的含硫香料

表 12-1 列出部分其他结构的含硫香料的化学结构和香味特征。

表 12-1　部分其他结构的含硫香料的化学结构和香味特征

名称	化学结构式	香味特征或应用建议
二甲基亚砜 （FEMA 号 3875）		脂肪香，油脂气，干酪香，咸味，蘑菇、大蒜香气[7]
硫代乙酸 （FEMA 号 4210）		可用于焙烤食品、饮料、肉制品、奶制品、调味品等香精[5]

续表

名称	化学结构式	香味特征或应用建议
4-甲基-4-(3′-呋喃基)-2-硫代戊酮		烘烤肉香味[8]
二巯基甲烷 (FEMA 号 4097)		可用于焙烤食品、肉制品等香精[5]
1-巯基-1-甲硫基乙烷		硫磺样、肉香、洋葱香[3,4,9,10]
1-巯基-1-甲硫基丙烷		硫磺样、烤香、韭葱香、洋葱香、肉香[4]
1-巯基-1-甲硫基-2-甲基丙烷		洋葱香、肉香[4]
1-巯基-1-甲硫基-3-甲基丁烷		洋葱香、肉香[4]
2,4-二巯基-3-硫杂戊烷		肉香、洋葱香、细香葱香
4-甲基-2,3,5-三硫杂己烷		硫磺样、洋葱香、肉香[4]
2,4,6-三甲基-1,3,5-二噻嗪		鸡肉香味[8]
2,4,6-三甲基-1,3,5-二氢二噻嗪 (FEMA 号 4018)		肉汤香、HVP样、硫磺样、橡胶样、吡嗪样、烤香、烤牛肉香、肉香[4,9,11~13]
2,6-二甲基-4-乙基-1,3,5-二氢二噻嗪		硫磺样、噻唑样、糖果样、烤香、肉香[4]
4,6-二甲基-2-乙基-1,3,5-二氢二噻嗪		煮蔬菜香、青香、豌豆样、大蒜香、韭葱香、烤洋葱香、烤花生香[4]

续表

名称	化学结构式	香味特征或应用建议
2,6-二甲基-4-丙基-1,3,5-二氢二噻嗪		甘蓝香、大蒜香、青香、蘑菇香[4]
4,6-二甲基-2-丙基-1,3,5-二氢二噻嗪		韭葱香、硫磺样、洋葱香、烤花生香、烤香、坚果香[4]
2,6-二甲基-4-异丙基-1,3,5-二氢二噻嗪（FEMA 号 3782）		硫磺样、青香、枯茗样、肉香、烤香[1,4]
4,6-二甲基-2-异丙基-1,3,5-二氢二噻嗪（FEMA 号 3782）		韭葱香、坚果香、洋葱香、烤香、烤花生香[1,4]
2,6-二甲基-4-异丁基-1,3,5-二氢二噻嗪（FEMA 号 3781）		热带水果香韵、果香、青香、烤香、可可香[1,4]
4,6-二甲基-2-异丁基-1,3,5-二氢二噻嗪（FEMA 号 3781）		硫磺样、橡胶样、细香葱香、甘蓝香、烤花生香、爆玉米花香、脂肪香、烤香[1,4]
6-甲基-2-乙基-4-异丁基-1,3,5-二氢二噻嗪		甜的、韭葱香、洋葱香、大蒜香[4]
6-甲基-4-乙基-2-异丁基-1,3,5-二氢二噻嗪		苦的、焦糊、烤香[4]

名称	化学结构式	香味特征或应用建议
6-甲基-2-异丙基-4-异丁基-1,3,5-二氢二噻嗪		蘑菇香、麦芽香[4]
6-甲基-4-异丙基-2-异丁基-1,3,5-二氢二噻嗪		硫磺样、脂肪香、药草香[4]
2-甲基-6-异丙基-4-异丁基-1,3,5-二氢二噻嗪		硫磺样、橡胶样、焦糖香味、可可香[4]
6-甲基-2,4-二异丁基-1,3,5-二氢二噻嗪		硫磺样、油腻、焦糊、洋葱香、糖果香、烤香[4]
2,4,6-三异丁基-1,3,5-二氢二噻嗪（FEMA 号 4017）		可用于焙烤食品、肉制品、调味品等香精[13]
2,4-二甲基四氢-4H-吡咯并[2,1-d]-1,3,5-二噻嗪		硫磺样、脂肪香、焦糊、洋葱香、烤花生香、面包样、咖啡样、强烈的烤香[3,4]
4-甲基-2-乙基四氢-4H-吡咯并[2,1-d]-1,3,5-二噻嗪		脂肪香、洋葱香、肉香[3]
4-甲基-2-异丙基四氢-4H-吡咯并[2,1-d]-1,3,5-二噻嗪		脂肪香、酸败味、焦糊、烤香、烤洋葱样、韭葱香、烤花生香[3]
4-甲基-2-异丁基四氢-4H-吡咯并[2,1-d]-1,3,5-二噻嗪		橡胶样、烤香、脂肪香、焦糊[3]

续表

名称	化学结构式	香味特征或应用建议
2,4-二甲基六氢-4H-吡啶并[2,1-d]-1,3,5-二噻嗪		硫磺样、橡胶样、洋葱香、烤肉香气[3]
2,4,6-三甲基全氢化-1,3,5-㗁噻嗪		硫磺样、橡胶样、蘑菇样香气[3]
1,2,3,5,6-五硫杂环庚烷(香菇素)		香菇[14]、羊肉[15]挥发性香成分
1,3,5,7-四硫杂环辛烷		葡萄酒香成分[14]
环蒜氨酸		存在于洋葱中[14]
丙基(3,4-二甲基-2-噻吩基)二硫醚		油炸洋葱特征香味化合物[16]
异硫腈酸丁酯(FEMA号4082)		可用于焙烤食品、肉制品、奶制品、调味品等香精[5]
异硫腈酸-4-甲硫基-3-丁烯酯		萝卜特征香味化合物[16]
异硫腈酸-4-戊烯酯		辣根特征香味化合物[16]
异硫腈酸苯乙酯(FEMA号4014)		水芹特征香味化合物[13,16]
蒜素亚砜		大蒜特征香味化合物,洋葱、大蒜、肉、蔬菜香韵[16,17]
硫代丙磺酸丙酯		洋葱特征香味化合物[18]

续表

名称	化学结构式	香味特征或应用建议
4-羟基-5-甲基-3-二氢噻吩酮		烤肉香味[19]
乙基 4,5-二甲基-2,3-二氢-3-噁唑基二硫醚		肉香[8]
2,4-二甲基-2-巯甲基-3-噻唑啉		炸洋葱香、肉香[20]
2-乙酰胺基硫代乙酸乙酯（FEMA 号 4039）		可用于焙烤食品、奶酪、肉制品、调味品等香精[21]
		香豆素香、酚样[22]
		土豆香、硫胺素、牛奶样[22]
		硫磺样、肉香、花生香、豌豆香[22]
		蔬菜香、洋葱香、烤香[22]
		土豆香、硫磺样、肉香、甘蓝香[22]

参 考 文 献

[1] Smith R L, Ford R A. Recent progress in the consideration of flavoring ingredients under the food additives amendment, 16. GRAS Substances. Food Technology, 1993, 47(6): 104~117

[2] Hall R L, Oser B L. Recent progress in the consideration of flavoring ingredients under the food additives amendment, Ⅲ. GRAS Substances. Food Technology, 1965, 19(2): 151~197

[3] Werkhoff P, Brüning J, Emberger R et al. Flavor chemistry of meat volatiles: new results on flavor components from beef, pork, and chicken//Hopp R, Mori K. Recent Developments in Flavor and Fragrance Chemistry. New York: VCH Publisher, 1992. 183~213

[4] Werkhoff P, Brüning J, Emberger R et al. Studies on volatile sulphur-containing flavour components in yeast extract//Bhattacharyya S C, Sen N, Sethi K L. 11th International Congress of Essential Oils, Fragrances and Flavours. Proceedings: Volume 4, Chemistry Analysis and Structure. New Delhi: Oxford & Ibh publishing Co. Pvt. Ltd., 1989. 215~243

[5] Smith R L, Cohen J, Doull J et al. GRAS flavoring substances 22. Food Technology, 2005, 59(8):24~62

[6] 冯驸,胡卫兵,刘红霞.二苄基二硫醇的合成.湖北民族学院学报,2006,24(1):69~70

[7] 舒宏福.咸味香料(下).香料香精化妆品,2006,(1):22~26

[8] 黄荣初,王兴凤.肉类香味的合成香料.有机化学,1983,(3):175~179

[9] Macleod G. The flavor of beef//Shahidi F. Flavor of Meat and Meat products. New York: Blackie Academic & Professional, 1994. 4~33

[10] Werkhoff P, Brüning J, Emberger R et al. Isolation and Characterization of Volatile Sulfur-Containing Meat Flavor Components in Model Systems. J. Agric. Food Chem., 1990. 38(3):777~791

[11] Ashurst P R. Food Flavorings. Second edition. New York: Blackie Academic & Professional, 1995. 128~129

[12] Farkas P, Sadecka J, Kovac M et al. Key odorants of pressure-cooked hen meat. Food Chemistry, 1997, 60(4):617~620

[13] Smith R L, Doull J, Feron V J et al. GRAS flavoring substances 20. Food Technology, 2001, 55(12):34~55

[14] Morton I D, Macleod A J. Food Flavors, Part A. Introduction. New York: Elsevier Science Publishing Company Inc., 1982. 176~184

[15] Cramer D A. Chemical compounds implicated in lamb flavor. Food Technology, 1983, (5):249~257

[16] Boelens M H, Gemert L J V, Rowe D. Volatile character-impact sulfur compounds and their sensory properties. Perfumer & Flavorist, 1993, 18(3):29~39

[17] Rowe D J. Aroma chemicals for savory flavors. Perfumer & Flavorist, 1998, 23(4):9~16

[18] Flament I. Molecular gastronomy. Perfumer & Flavorist, 1997, 22(1):1~8

[19] Shaikh Y. Aroma chemicals in meat flavors. Perfumer & Flavorist, 1984, 9(3):49~52

[20] 黄小凤,李晓东,李中林.杂环类香料的现状与展望.化学通报,1995,(8):1~16

[21] Smith R L, Cohen S M, Doull J et al. GRAS flavoring substances 21. Food Technology, 2003, 57(5):46~59

[22] Güntert M, Bertram H J, Emberger R et al. Thermal degradation of thiamin(Vitamin B_1)//Mussinan C J, Keelan M E. Sulfur Compounds in Foods. Washington DC: American Chemical Society, 1994. 212~213